MANAGING WATER WELL DETERIORATION

Managing Water Well Deterioration

Robert McLaughlan

National Centre for Groundwater Management, University of Technology, Sydney, Australia

IAH-publication: International Contribution to Hydrogeology (ICH)
Volume 22

Taylor & Francis
Taylor & Francis Group

LONDON AND NEW YORK

Library of Congress Cataloging-in-Publication Data

McLaughlan, Robert, 1958–
    Managing water well deterioration / Robert McLaughlan.
        p. cm. – (International contributions to hydrogeology v. 22)
    Includes bibliographical references and index.
    ISBN 905809247X
        1. Wells–Fouling. 2. Wells–Corrosion. 3. Wells–Maintenance and repair. I. Title II.
Series.

TD405 .M38 2002
628.1′14–dc21

2001056546

Published by Taylor & Francis
2 Park Square, Milton Park, Abingdon, Oxon, OX14 4RN
270 Madison Ave, New York NY 10016

Transferred to Digital Printing 2006

Cover design: Studio Jan de Boer, Amsterdam, The Netherlands.
Typesetting: Charon Tec Pvt. Ltd, Chennai, India.

ISBN 90 5809 247 X

Publisher's Note
The publisher has gone to great lengths to ensure the quality of this reprint
but points out that some imperfections in the original may be apparent
Printed and bound by CPI Antony Rowe, Eastbourne

Contents

List of Figures

List of Tables

Preface

The deterioration of water wells due to fouling and corrosion is a global problem. The increased demands placed upon our aquifer systems are causing a much greater awareness and possibly occurrence of water well deterioration within our existing water wells.

Water wells that are used for groundwater extraction, injection and remediation can all be impacted by well deterioration. To varying degrees, all these water wells disturb the microbial, hydrochemical and hydraulic environments that exist within an aquifer. Groundwater extraction wells with long well screens cause mixing of groundwater from different parts of an aquifer, which may then react and cause fouling. Groundwater injection wells may introduce incompatible water chemistries or particulates to an aquifer leading to well fouling or aquifer plugging. Even the construction of water wells will introduce into the aquifer a foreign material that may be reactive and corrode. All of these interactions lead to the deterioration of water wells. It is just that in some environments it is at a much greater rate than in others.

The management of a water well system can be a complex task. Groundwater wells and aquifers are often valuable assets. Water wells may have to meet specific performance indicators related to water yield, water quality, operational costs and asset life. Fouling and corrosion processes can impact upon all of these performance indicators. In such environments it is necessary to have an explicit well maintenance strategy. An efficient and effective strategy involves finding the right mix of preventative and corrective maintenance activities. The challenge is to be able to do this in an environment where equipment failure behaviour is poorly known and the cost of various maintenance activities to achieve a given level of water well performance is difficult to estimate.

Managing Water Well Deterioration fills a need within the literature for an academically based informative text that incorporates practical advice. It contributes to the emerging knowledge base about water well deterioration. There are a number of other useful papers and books that deal with fouling and to a much lesser extent corrosion of water wells from a scientific and descriptively orientated framework. There is now a need to embed these understandings of the scientific basis of water well deterioration processes within management frameworks to provide a comprehensive approach to dealing with water well deterioration. How this is done will often reflect local judgement, customs and practices, available resources as well as understanding of the well deterioration processes that operate at a particular site. It is hoped that the frameworks for problem solving and the knowledge enabling the management of these problems that is presented in this book can usefully contribute to the development of strategies to manage water well deterioration.

Acknowledgements

This book has a long history with support from many sources. The background studies for this work were carried out as part of a project into the Deterioration of Bore Performance, funded by the Australian Research Advisory Council (subsequently the Land and Water Resources Research Development Corporation) and all Australian mainland Water Agencies. Dr Richard Stuetz is thanked for his contribution to the previous project and for the microbiology photographs. The field assistance of the Water Agencies, State Electricity Commission, Geelong and District Water Board, Bundaberg City Council, Gatton Shire Council and Gatton Agricultural College during previous investigations is gratefully acknowledged. BHP and Sandvik are thanked for providing the corrosion coupons without which the corrosion work could not have been completed. Graeme Kelly of Corrotec Services is thanked for his considerable assistance in the corrosion work. Funding from the Land and Water Resources Research Development Corporation allowed various technical monographs to be produced and is gratefully acknowledged. Special thanks are given to Lewis Clark from Clark Consult Ltd for his review which identified a number of ways the book could be improved. Ian Simmers from the International Association of Hydrogeologists is thanked for his contribution to getting this book published. Thanks also to Julie for her patience whilst the book was in the long process of being 'finally' finished.

1 PROCESSES

Within a water well system there are a number of different processes acting, which can cause or accelerate the deterioration of well performance through fouling or corrosion. Fouling can occur due to the formation of particulate, mineral scale or biofouling deposits. Corrosive forces operating on groundwater wells can involve both erosion and electrochemical reactions, with both plastics and metals that are commonly used in well construction affected.

1.1 Fouling

It is now widely recognised that microbes exist in most if not all groundwater environments. Many water supply wells are located in shallow groundwater systems where there are potentially very active and robust microbial communities. This has often been demonstrated by the success of using native bacterial populations to bioremediate organic groundwater contamination. In fact there has been little success in establishing introduced bacteria into aquifers as they tend to be out competed by the native populations. This suggests that when biofouling problems occur in wells after a number of years of trouble free service then it is unlikely to be that 'iron bacteria' have been introduced to the well. It is more likely that the symptoms of fouling have not been apparent before or that the environmental conditions within the aquifer have changed which has allowed the bacteria to flourish.

Increasingly it is being understood that bacteria play a key role in controlling many of the chemical reactions that occur within groundwater. This is particularly the case with many of the dissolved metals in groundwater (Fe and Mn) that make up a significant component of all well fouling deposits. It is often not practical from a well management perspective to distinguish whether dissolved iron (Fe^{2+}) is converted to the oxidised form Fe^{3+} abiotically or microbially since the bacteria that carry out this reaction tend to live under the same environmental conditions where the iron oxidation would occur chemically. The types of deposits where bacteria play a role in deposit formation are called biofouling deposits. These deposits are the most common well fouling deposits.

Particulate and mineral scale deposits also occur within groundwater wells but are much less widespread. Mineral scale is precipitated due to chemical reactions and does not contain any microbiological matter. These deposits may be formed on a surface from crystallisation or be comprised of loosely adhered particles. Particulate deposits are formed from the transport and deposition of aquifer material. The material making up these deposits generally consists of sand or silt sized particles.

1.1.1 Biofouling

The range of micro-organisms which inhabit the subsurface include moulds, fungi and bacteria however with regard to well fouling the principal fauna involved are bacteria. The bulk of these bacteria live attached to surfaces in the aquifer but at various times in their life they may also be mobile. These attached bacteria live in a biological film (biofilm), which can comprise both aerobic and anaerobic environments and include a diverse microbial community. When a biofilm accumulates it becomes a biofouling deposit. Biofouling deposit formation involves the bacterial production of extracellular polymers (ECP) and the subsequent accumulation of various inorganic compounds and particles. The ECP serves a variety of purposes for the bacteria including adhesion, nutrient collection and buffering against environmental change. The accumulation of inorganic components in the biofilm involve both active and passive processes. The active processes include autotrophic bacteria such as *Gallionella* (Lutters-Czekalla, 1990) utilising iron for growth and heterotrophic organisms using the organic component of a metal–organic chelate (McCrae *et al.*, 1973). The passive processes involve the formation and adhesion of particles (e.g. Fe, Al, carbonates and clay) to the ECP as water passes through or across the biofilm surface.

The net rate of biofilm deposit accumulation is a function of bacterial activity, particle availability and biofilm shear forces.

These factors are dependent upon:
• Bacterial activity
    – Nutrient availability
    – ECP production
• Particle availability
    – Volume of flow past a surface
    – Precipitation mechanisms such as CO_2 degassing, pH changes and oxidation
    – Aquifer composition (e.g. clay)
• Biofilm shear forces
    – Flow rate, turbulence

1.1.2 Mineral Scaling

Mineral scale deposition within groundwater supply wells may occur due to the mixing of incompatible waters and/or changes in groundwater temperature or pressure during pumping.

Within pumping wells the mixing of incompatible waters is the most likely primary process causing the buildup of mineral scale. Groundwater wells may penetrate several different groundwater chemistries either within the same or different geological strata. These different waters can be brought together by screening the well through the different layers or through parting or corrosion of the casing allowing different waters to mix within the well. If these waters are incompatible such as a carbonate rich water mixed with high salinity water (high calcium) then a rapid accumulation of mineral scale can occur (Section 6.6).

Changes in the water temperature and/or pressure of carbon dioxide in groundwater during entry to the well and pumpage to the surface can cause degassing of CO_2. In response to changes in CO_2 concentration in the groundwater other chemical reactions occur. There can be a tendency for mineral scale to precipitate. This process is more likely to be a secondary cause of mineral scaling in shallow water supply wells. In wells with higher groundwater temperatures and greater CO_2 concentrations it can be more important. This process can operate in conjunction with mineral scaling caused by the mixing of incompatible waters.

1.1.3 Particulate Fouling

The particles within an unconsolidated aquifer comprise both load bearing sands and interstitial fines. Particulate fouling within the aquifer involves the buildup of fine particles close to the well in sand bridges, which can reduce the porosity of the aquifer. The mobilisation of fines within the aquifer can lead to their entry into the well causing particulate deposits and/or pump corrosion.

The principal factors involved in particle accumulation within and around groundwater wells include:
- Poor well design
    - Screen selection
    - Screen placement
- Inadequate well development
    - Contamination by drilling mud
    - Gravel pack fines
    - Formation fines
- Operational factors
    - Screen dewatering
    - Screen entrance velocities
    - Pump cycling
- Injected water quality

1.1.3.1 Well Design
If the design of the well is not correctly matched to the physical and chemical properties of the aquifer then particle accumulation and migration can occur. A well screen which has straight cut, punched or gauze type openings can become clogged through the bridging of elongate particles while the V-shaped slots used in continuous slot screens which widen inwardly are less prone to clogging. A metal screen may produce corrosion products that cement fine particles together, lowering the area of the screen available for groundwater flow. Alternately, corrosion of the slots will allow particle entry to the well.

Sand entry may also occur due to incorrect screen mesh selection especially in aquifers with a large variation in aquifer particle size. A well screen located in a coarse-grained aquifer near the margin of a finer-grained aquifer may clog or allow sand entry through the entrainment of finer material from the adjacent aquifer. Within longer well screens there may be significant variations in the screen entrance velocities due to differences in aquifer permeability. Even within homogeneous aquifers there is evidence to suggest that there are upward flow velocity losses, which may cause non-uniform screen entrance velocities. These effects were found to decrease as the casing diameter increased from 150 to 200 mm (Howsam *et al.*, 1995). This can impact upon sand production by mobilising fines in the sections of casing with higher screen entrance velocities.

1.1.3.2 Well Development
Well development techniques are designed to remove fines from the immediate vicinity of the well. During well construction fine particles will be introduced into and moved around the drilled hole and surrounding aquifer through the circulation of drill cuttings or drilling mud. Drilling through an interbedded unconsolidated aquifer will cause a mixing of the sediments causing a reduction in the porosity near the well. The addition of drilling mud or the hydration of clays within the formation can cause the clogging of pore spaces within the aquifer. Drilling mud and lost circulation fluids often contain fines, which may invade the aquifer. Drilling mud that contains biodegradable components may not break down as quickly as expected and their removal as much as possible by mechanical means or chemically enhanced breakdown should be considered. Fine material if present within the gravel pack can provide a source of material for clogging. Only clean inert gravel (e.g. quartz) should be used as lateritic and calcareous gravel may decompose in water or during subsequent chemical well treatment if it is required.

During well development the removal of fines from the aquifer creates sand bridges that are stable up to the well development flow rate used. If the flow rate during pump operation is greater than the well development flow rate the bridges may collapse resulting in sand production until new bridges are formed. If the flow rate is constantly varied then no stable bridging may occur and sand ingression into the well would be expected. It has been found in practice that a well may make a uniform amount of sand/fines independent of the pumping rate until a critical flow rate is reached. Production above this rate causes increasing amounts of sand production. The acceptable sand content will depend on groundwater extraction system design, end use of water and well rehabilitation costs. There are no standards set for the maximum allowable sand content of pumped water, however a practical limit (for pump wear) is often 5–8 mg/L. Suggested sand content limits range from 1 mg/L for waters directly in contact with food or beverages to 5 mg/L for domestic and industrial users with up to 15 mg/L for flood type irrigation (EPA, 1975).

1.1.3.3 Well Operation
Pumping a well at flow rates greater than the design or well development yield will increase screen entrance velocities which can allow finer particles to be mobilised

within the aquifer and enter the well. Overpumping may also cause dewatering of the upper well screens, which enhances chemical precipitation and the cementation of finer particles at the well screen. Intermittent pumping of the well can break down particle bridges established during continuous pumping and mobilise fine particles.

1.1.3.4 Injected Water Quality

An accumulation of particles immediately outside the well screen either within the gravel pack or in the aquifer will result from the injection of sediment-laden water into recharge wells. The distance the particles penetrate into the aquifer is dependent upon the size of the suspended sediment relative to the size of the aquifer pores. If the particles are relatively large then particle straining occurs through the mechanical removal of particles due to restrictions in the aquifer pores. Finer particles may then accumulate within these particle or sand bridges. These bridges will occur near the gravel pack–aquifer interface and can often be removed by backwashing. Particles that are small relative to the aquifer pore size are capable of travelling further into the aquifer and may be immobilised by physico-chemical forces (McDowell-Boyer *et al.*, 1986). It is possible that re-orientation of sand/mica particles within an aquifer due to the injected flow can also lead to a well yield problem. Incompatibilities in water chemistry between the injected and aquifer water chemistry may also cause the formation of mineral scale.

1.2 Corrosion

The corrosion of metals may be caused by electrochemical or by physical processes through particle or fluid impact. The electrochemical reactions can be generated from contact between system components with groundwater or stray electrical currents and be affected by the presence of bacteria. Plastic well components can be affected by the presence of certain organic compounds in the groundwater.

1.2.1 Metal

In groundwater wells the corrosion of active metals (e.g. steel) is generally an electrochemical mechanism that involves the transfer of electrons. The electrons are released from the active metal when it changes its ionic status. In the case of iron this occurs when elemental iron (Fe^0) changes to the dissolved iron (Fe^{2+}) form. The corrosivity of various metals reflects their different tendencies to form ions and be dissolved in groundwater. Where the metal is corroded and electrons are generated is called the anode. The electrons flow through the metal to the surface where the cathode is formed (Fig. 1.1). An electric current will then flow through the groundwater via dissolved ions back to the anodic surface. The ability of the water to pass the current is measured by its conductivity, which reflects the amount of dissolved ions in the water. In this way a circuit or corrosion cell is formed and the rate of corrosion is proportional to the rate at which electrons are accepted at the cathode that is equal to

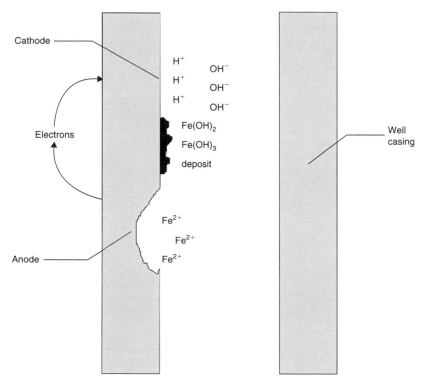

Figure 1.1 Corrosion processes on metal.

the rate of current flow through the water. The rate of corrosion is limited by polarisation of the metal surface. Polarisation can occur through the buildup of surface deposits. For steel casing the dissolved iron created at the anode can form an iron hydroxide deposit (rust), which helps polarise the surface. A more extensive review of electrochemical corrosion is given in Appendix C.

The size of the corrosion cell may be small (localised pitting) or large (uniform corrosion). In groundwater well casing, localised pitting is often a more severe type of corrosion since this may rapidly penetrate the casing and allow material from the aquifer into the well. Anodic and cathodic areas may be dynamic and shift causing a more uniform rate of corrosion across the metal surface.

1.2.1.1 Biocorrosion

Micro-organisms often live in a biofilm, which can contain a range of chemical environments that do not occur in the bulk water chemistry. These environments can favour electrochemical corrosive processes, which may not occur in the rest of the groundwater. The major microbial influence on corrosion in water wells occurs from concentration cell formation and cathodic depolarisation (Appendix C).

1.2.1.2 Stray and Induced Currents

Current flow and hence corrosion of metal surfaces can be increased by the imposition of stray or induced electrical currents. These currents occur from sources outside the well casing and include the grounding of electrical equipment such as pumps onto the wellhead or the electrical field from high power transmission lines inducing currents in pipelines. Corrosion occurs at the anode where the current leaves the metal surface to the soil or groundwater.

1.2.1.3 Erosion

Corrosion through erosion involves the mechanical removal of protective layers such as iron oxides and carbonate films that can then lead to corrosion. This type of corrosion often appears above a critical flow rate particularly where there is a constriction of flow or change in flow direction. Erosion from particles may occur above a critical particle concentration so that corrosion occurs as a result of the sum of individual particle impacts.

Corrosion from cavitation involves the formation and collapse of gas bubbles. If the water pressure is low where the water enters the pump inlet then the water vaporises creating vapour pockets. As water flows through the pump it can be subjected to higher pressures and the vapour implodes. When this occurs against a solid surface the localised pressure change can damage the metal surface or remove protective surface films. The roughened surface can then provide sites for further bubble formation.

1.2.1.4 Environmental Factors

The environment around a metal surface will control the extent of the corrosion process. Both the chemical and physical properties of water can be significant:

- Salinity

The corrosion rate generally increases as the salinity of the water increases. The high conductivity water will increase the current transfer between the different sections of the casing or between the casing and the earth. Some metal salts (e.g. Cl^-, SO_4^{2-}) forming acids (e.g. HCl, H_2SO_4) in water are involved in pitting corrosion, while ions such as bicarbonate, carbonate and hydroxide may decrease the corrosion rate by forming protective scales.

- Oxygen

The presence of oxygen is recognised as the major factor influencing corrosion in most wells. For the submerged sections of casing the oxygen will be present dissolved in water. The oxygen depolarises the cathode, which increases corrosion. The amount of oxygen supplied to the metal surface is a function of the water temperature, flow rate and the presence of a scale.

- Carbon dioxide

Carbon dioxide is soluble in water and reacts to form a weak acid (H_2CO_3). Corrosion by carbon dioxide occurs in groundwater wells particularly in deep wells where its effect is exacerbated by elevated water temperatures. The corrosion is often very localised in the form of pits, gutters or attached areas with abrupt changes from corroded to non-corroded areas.

- Hydrogen sulphide

Hydrogen sulphide is recognised as a corrosive agent in the natural gas industry where the water has high temperatures, pressures and chloride content. These conditions are unlikely to occur in many water supply wells. However the sulphide inclusions particularly in older casing when quality control procedures were not rigorous is possible. These can act as sources of pit nucleation.

- Flow rate

Flow dependent corrosion may be classified as chemical or mechanical (erosion). In general, corrosion rates increase with increasing velocities up to a point. This reflects the increased reductant (e.g. oxygen, hydrogen) supply to the reaction surface (cathode). At high velocities any scale present can become protective of the surface.

- pH

The relationship between pH and corrosion rate reflects a mixture of hydrogen ion (H^+) effects and carbonate equilibrium processes. The increased H^+ concentration that occurs at a low pH accelerates the corrosion of most metals. At a high pH there are more carbonate and hydroxide ions that can increase the tendency of the water to form scales which protect the metal surfaces.

- Temperature

Elevated water temperatures in conjunction with other environmental factors to increase corrosion rates.

1.2.2 Plastics

Although not strictly corrosion, the synthetic materials used for well construction, sampling and pumping of groundwater contaminated with organic compounds can be susceptible to structural degradation. These degradation processes may be oxidative, mechanical, microbial and chemical. Plastics may primarily be degraded by solvation where particular chemicals penetrate the plastic causing swelling and softening. These changes lead to a change in the material properties that can cause structural failure. Particularly vulnerable are well screens since they often contact the highest contaminant concentrations in the aquifer and well pump components made of butyl rubber. Compilations of material compatibility tables are available (McCaulou *et al.*, 1995) to assist the design of wells in these environments.

2 PROBLEMS IN DIFFERENT TYPES OF WELLS

Wells are used in a range of applications including water supply, groundwater remediation and subsurface injection. Certain well deterioration processes and well performance problems can be associated with the various applications of well technology.

2.1 Water Supply Wells

2.1.1 Fouling

The precipitation of chemical compounds from groundwater requires the presence of suitable concentrations of particular chemical species. There are several common hydrogeological environments where these precipitation reactions can occur.

These involve:
- Mixing of incompatible chemical species
- Mixing of chemical species which then involve a redox change of an element causing precipitation

The chemical species present in the groundwater will determine which of these two precipitation processes are dominant. Chemical species that are ready to precipitate when mixed with other suitable compounds include Al^{3+}, S^{2-}, CO_3^{2-}, SO_4^{2-}, Ca^{2+} and hydroxide ions (OH^-). Saline waters containing high concentrations of Ca^{2+} and Mg^{2+} ions when mixed with a carbonate (CO_3^{2-}) rich water will form a carbonate deposit (Fig. 2.1). This can occur when the well is screened in the saline water or if corrosion of the upper casing has allowed its entry into the well. This occurs in the case study presented in Section 6.6.

Precipitation may also occur where the hydrochemistry of the different layers within an aquifer have significantly different pH. In low pH waters Fe^{2+} and Al^{3+} are more likely to be present, since they are more soluble under these conditions. The geological conditions for low pH waters are often found in organic rich sediments. If this water is mixed with higher pH water from the other sections of the aquifer the solubility of these elements are decreased and precipitation occurs (Fig. 2.2). This occurs in the case studies presented in Sections 6.7 and 6.9.

The most common elements that need a change in ionic form through redox processes to precipitate are iron and manganese. The most common causes of redox changes are oxygen and bacteria. This scenario for deposit formation is the most common cause of iron biofouling in wells. In uncontaminated groundwater the waters near the water

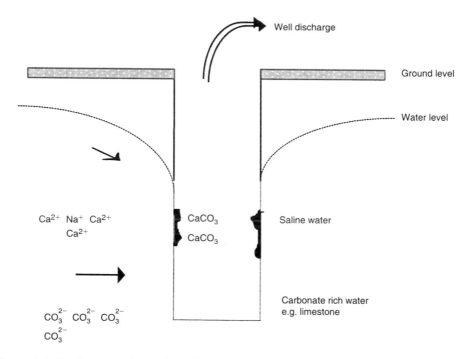

Figure 2.1 Fouling caused by mixing incompatible chemical species involving carbonates.

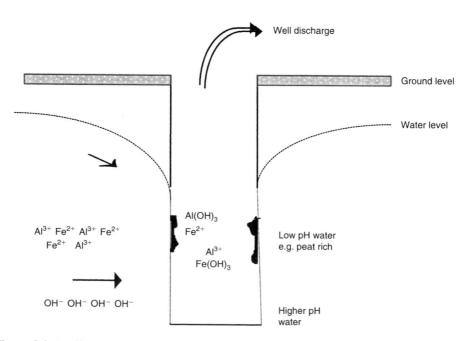

Figure 2.2 Fouling caused by mixing incompatible chemical species involving low pH water.

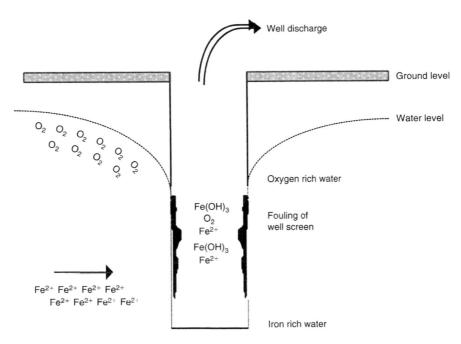

Figure 2.3 Fouling involving redox changes of chemical species.

table are enriched in dissolved oxygen (O_2) forming oxidised water. The groundwater at depth is often deprived of oxygen and is therefore more reduced. Under reduced conditions iron is more likely to be found in the dissolved form (Fe^{2+}). When these two waters are mixed the oxidising water aided by bacteria causes Fe^{3+} to occur which being insoluble forms a deposit with hydroxide ions (Fig. 2.3). This biofouling process occurs in some of the wells examined in the case study reported in Section 6.8.

The chemistry of some common chemical species associated with well fouling is discussed further in Appendix A.

Biofouling has been found to occur in discrete zones (McLaughlan *et al.*, 1993; Tyrrel & Howsam, 1990) within groundwater extraction systems. This may be where the chemical/microbial processes inducing fouling occur however these reactions are often not instantaneous and accumulation may occur elsewhere depending on the groundwater flow in the well system and microbial growth.

The possible occurrence of biofouling deposits within a groundwater extraction system can be divided into six zones (Fig. 2.4):
- Zone 1 – near the bottom of the screen
- Zone 2 – an isolated section of the well screen separated biofouling near the top of the screen

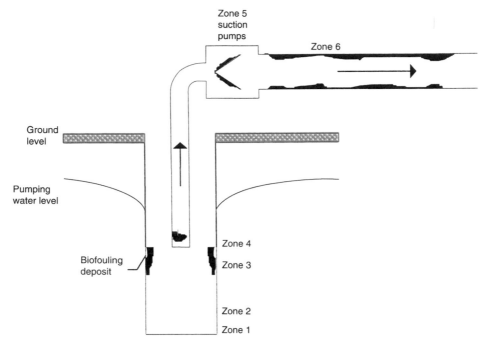

Figure 2.4 Iron biofouling zones in a groundwater extraction system.

- Zone 3 – the top section of the well screen
- Zone 4 – pump/suction inlet
- Zone 5 – around the surface pump impellers (this zone occurs in suction lift centrifugal pumps and not in submersible pumps)
- Zone 6 – the discharge side of the pump including well column pipes, degassing tanks and pipelines

Based on a study of over 40 wells the following observations of iron biofouling accumulation were made (McLaughlan *et al.*, 1993):

Zone 1
Deposits were never observed at the bottom of the well screen. When the bottom of the screen was isolated and pumped the water was initially often turbid compared with flow from further up the well screen. It is likely that the bottom screens in some wells produce very little flow. Generally in this environment aerobic heterotrophic bacterial activity was low and little groundwater mixing occurred leading to a low rate of biofouling. At a couple of sites biofouling did extend to within a couple of metres of the base of the well screen. These wells were constructed in a consolidated formation with no gravel pack present so an annular space existed between the well casing and the formation. Groundwater mixing could then occur to a significant extent outside the well casing. In all other wells the bulk of mixing probably occurred within the well.

Zone 2

No biofouling was ever observed in this zone probably because the most significant organisms observed in this study were aerobic–microaerobic bacteria and the preferred growth environment increases towards the top of the well. Fouling deposits due to particles, pH changes and carbonates may however occur in this zone.

Zone 3

If biofouling occurred within a well then luxuriant growths often occurred in this zone. In an unconfined aquifer this is the zone where the water entering the well is most likely to be a mixture of oxidised and reduced waters that the iron autotrophic and heterotrophic bacteria prefer. Each type of bacteria has their optimal environment in the oxidising (aerobic) zone and oxidising–reducing (aerobic–anaerobic) interface. The availability of particles for attachment to biofilms is greater in this zone since all groundwater must flow through this zone to the pump inlet.

Zone 4

Many groundwater systems are operated so that there is a minimum distance between the pumping water level and the pump inlet. In this region there is often the heaviest biofouling. Often the pump inlet is close to the top of the screen so that zones 3 and 4 coincide. Around zone 4 there can be increased oxygenation from atmospheric diffusion into the pumping water level surface within the well. The turbulence in this zone can also lead to CO_2 degassing. Both turbulence and oxygenation increase the availability of iron hydroxide particles. Around the submersible pump motor, biofouling was also significant and may be due to increased microbial activity related to heat from the pump motor.

Zone 5

In suction lift pumps the pump impellers are at the surface and significant changes in pressure occur in this region. Deposits can form on the impeller surface even though hydrodynamic forces causing biofilm shearing may be high.

Zone 6

In biofouled systems there was always biofouling in this zone. The groundwater within the well is generally under unstable conditions due to incomplete mixing of water from different depths. After the water has passed across the pump impeller it is better-mixed and more likely to precipitate particles which then increase the particle availability. The flow velocities are lower than in the pump column due to larger pipe diameters and so the shear forces removing the biofilm are reduced. At the suction lift centrifugal pump sites it was noted that biofouling was always greater on the discharge side rather than the suction side of surface mounted pumps.

These observations were made on wells where redox induced fouling was dominant. Although all iron biofouling wells may not conform to this distribution of deposits it is worthwhile to concentrate well remediation activity on the critical zones identified (zone 3). Ideally a closed circuit television (CCTV) log would be used to identify the zones of greatest fouling within a particular well before rehabilitation occurred.

These biofouling deposits can buildup on the pump motor (Plate 1) and on the well screen (Plate 2).

The distance that biofouling extends into the aquifer outside the well screen has not been established from field studies. Since biofouling relies on the mixing of different waters containing the essential components for fouling the limit of fouling will be the zone of groundwater mixing within the aquifer. This will vary based on well construction, hydrogeology and well hydraulics. If there is a thick gravel pack around the well there will be a greater opportunity for vertical mixing further from the well compared with a thin gravel pack or a naturally developed well. It is likely that the bulk of biofouling will occur on the screen and in some cases may extend out to the gravel pack/aquifer interface (van Beek, 1984). This is an important consideration when designing a well rehabilitation programme. The amount of chemical used and the development technique chosen may be different for removing fouling within the well compared with treating the gravel pack and aquifer.

2.1.2 Corrosion

The diversity of well construction types and groundwater chemistries make it difficult to be specific about what corrosion processes may occur in a particular well. There are specific parts of groundwater extraction systems (Roscoe-Moss, 1990) that are consistently found to have problems (Fig. 2.5). The rates of corrosion associated with each process and the various parts will vary significantly.

2.1.2.1 External Surface Casing
The corrosion of the external surface of well casing can be due to:
- Electrical currents
- Attack by groundwater
- Bacterially generated H_2S

Electrical currents can be generated at the wellhead if different metals are used in the discharge head, ancillary devices (e.g. flowmeter) and the pipeline. Stray currents can be induced onto flowlines from high voltage power transmission lines, cathodic protection groundbeds and electrical grounding systems. If these currents are significant they may be detected locally by connecting a voltmeter across the different surfaces. Stray current corrosion may be evidenced by the formation of isolated pits often in a line along the surface.

Potential currents can also occur underground between different parts of the casing. It is caused by the electrical properties of the various geological strata, the casing contacts (Plate 3). An oxygen concentration cell could occur at the casing contact between open porous strata (i.e. aerobic) and tight clay strata (i.e. anaerobic). In some strata (e.g. 'corrosive clays') sulphate reducing bacteria produce H_2S, which can corrode steel casing. These bacteria require anaerobic conditions although they survive in

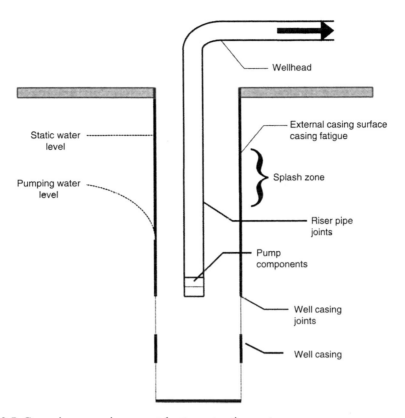

Figure 2.5 Corrosive zones in a groundwater extraction system.

environments with low oxygen concentrations. Saline water within a soil can also cause severe corrosion of metal surfaces. Most of the properties of groundwater that affect corrosion (Section 1.2.1.4) are also relevant to soils that have pore water. Various guidelines are available for a range of materials commonly used in pipelines (Prevost, 1987). Well constructions that include a sulphate resistant cement grout will insulate the casing from the strata.

2.1.2.2 Splash Zone

Corrosion can often occur in the zone between the static and pumping water levels. When the pump is turned on the water draws down to the pumping level leaving an oxygenated wetted surface on the casing. This layer creates a corrosive environment until the water film dries. Around the pumping water level the corrosion is often greatest due to the constant wetting from water level fluctuations. The corrosion rate over the rest of the splash zone is related to the frequency of casing wetting (Dunn *et al*., 2000).

2.1.2.3 Submerged Casing

The corrosion rate below the pumping level is variable due to the range of water chemistries that can occur. In wells subject to biofouling there will be deposit buildup on the well casing near the pump and the upper well screen. This area would then be subject to differential aeration cells and microbial corrosion. Welded joints between the casing and the well screen can be subject to galvanic corrosion because of the different compositions and metallurgical properties of the casing material, well screens and welds. Particularly vulnerable are small sections of a less resistive material (e.g. steel) placed next to larger sections of resistive casing (e.g. stainless steel). These small sections of casing become anodic to the well screen leading to corrosion of the casing.

Well casing, screens and riser pipe are subject to various stresses during installation and well operation. There are axial forces that can pull the casing apart or compress it and radial forces that will tend to collapse the casing. Casing fatigue can occur from inadequate well design by suspending or having long lengths of unsupported casing in an open hole. During well construction the casing may be jacked, driven or twisted in an effort to make the casing fit the hole. The stresses imposed on the casing can cause it to deform and weaken the casing joints. The sections of fatigued casing are also more vulnerable to corrosion processes.

If the casing failure is very localised then the performance of the well may be partially impaired. Small amounts of sand may enter the well increasing pump corrosion. However if the thickness of the casing is reduced over a larger area then the risk of massive failure such as well casing rupture and loss of the pump down the well increase. The relationship between collapse strength and casing thickness is complex. For a 150 and 200 mm casing a reduction in the casing thickness of 25% and 45% result in a reduction in the collapse strength of about 45% and 70% respectively. How important this is in a particular well depends on the stresses imposed by the water and the strata as well as the safety margins incorporated into the well design. Detailed information on casing stresses is provided by Roscoe-Moss (1990) while Ireland (1978) has conducted an extensive survey of casing failure.

2.1.2.4 Pump Components

The pump components include both downhole pump items as well as the riser pipe used to transmit the pumped water to the surface. The threaded joints of metal riser pipe are vulnerable to several types of corrosion. When riser pipes are screwed together the tools used to grip the pipe often roughen up the pipe surface near the joints. Any protective coating on the pipe may be damaged. These unprotected areas then become anodic and corrosion can be initiated. The threaded joints are also liable to crevice corrosion from corrosive groundwater if they are not sealed with a hard setting jointing compound (Plate 4). The water under pressure in the riser pipe can erode the joint further as it discharges into the well casing (Plate 5). In the riser pipe used for line shaft pumps the shaft centralisers can cause flow restrictions resulting in flow velocities in excess of 20 m/s, which can initiate erosion.

The bearings in submersible pumps are liable to failure from sand pumping particularly as many impellers are free to float rather than being fixed axially to the pump shaft.

The pumps are designed to operate in down thrust with the bearing clearance sufficient to be lubricated by the pumped fluid but exclude abrasive particles. If there is no load during pump startup the impellers go into up thrust causing the thrust bearing clearance to increase allowing very large grains to enter. When the pump stabilises the impellers go into down thrust with abrasion of the thrust bearing from the sand grains (Wilson, 1990).

The surface of pump impellers can be vulnerable to erosion from particles as well as cavitation. The surface of materials subjected to erosion is characterised by grooves and gullies that are typically undercut in the direction of flow (Plate 6). Pump performance will be impacted by wear on either the leading edge or the outside edge of the impeller. Pumped water can be re-directed through the increased impeller clearance resulting in it being pumped again. This leakage reduces the efficiency of the pump. Wear on the impeller surface can also occur through cavitation. This is evidenced at the wellhead by noisy pump operation and fluctuations in both power consumption and water yield.

2.2 Remediation Wells

There has been an increase in the use of remediation wells in an effort to minimise the potential impact of contamination on groundwater resources. Wells are used for both the monitoring and recovery of contaminated groundwater.

There are many different scenarios for fouling to occur given the variability in hydrogeology and hydrochemistry at contaminated sites. At many petroleum hydrocarbon and landfill sites, micro-organisms rely on degrading these contaminants by oxidising this organic matter and other reduced compounds for growth. This requires the transfer of electrons from an electron donor to an electron acceptor and is known as the terminal electron acceptor process (TEAP). The most common TEAPs in groundwater environments are oxygen reduction, nitrate reduction, manganese reduction, iron reduction, sulphate reduction and carbon dioxide reduction (methanogenesis). Since the bacteria carry out those reactions that are most energetically favourable first, there is a sequence to the reactions. Oxygen is utilised first followed by nitrate, manganese, iron, sulphate and carbon dioxide. When microbially degradable organic matter (e.g. petroleum) is released near the water table, specific hydrochemical zones occur within the aquifer (Fig. 2.6). Due to the high contaminant concentrations near the source, all of the electron acceptors except carbon dioxide may have been consumed, resulting in methanogenesis. At the fringes of the plume where contaminant concentrations are much lower then oxygen reduction may be the dominant process occurring. The boundaries between the hydrochemical zones are a gradation rather than abrupt change in chemistry and the size of the individual zones vary at different sites. These zones are significant because byproducts of some of these metabolic reactions are dissolved ions, which can subsequently cause fouling both of the aquifer and of an extraction or monitoring well. Iron reduction will produce Fe^{2+}, manganese reduction (Mn^{2+}) and sulphate reduction (S^{2-}). The production of organic acids during microbial

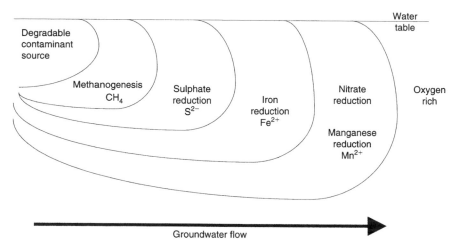

Figure 2.6 Hydrochemical zones in an organic contaminated aquifer.

respiration may also lower the pH that can increase the concentration of certain ions (e.g. Al^{3+}, Ca^{2+}, Mg^{2+}).

The amount of well fouling experienced in remediation wells will be impacted by their location within these zones and the amount of mixing with uncontaminated groundwater in the well. The degree of plastic corrosion (e.g. uPVC) is largely controlled by the concentration of the contaminant if it is a solvent of the particular plastic (McCaulou *et al.*, 1995).

2.2.1 Monitoring Wells

Within monitoring wells the buildup of fouling deposits can significantly impact measured water quality parameters such as contaminant concentrations and turbidity. Deposits will concentrate particular metals which when suddenly released can give values not representative of the groundwater which the well is designed to sample.

At a site contaminated with chlorinated hydrocarbons the monitoring well casing was found to have a black deposit (Plate 7) covering 50% of the uPVC casing below the water table. The deposit comprised Si, Al, Fe and S as quartz, greigite and elemental sulphur. It was estimated that if these deposits on the casing were remobilised during sampling with a bailer then the sulphide levels in the well would have increased by 29 mg/L. There was also a deposit 5 cm deep in the bottom of the well casing. Since sulphide was being used as indicator of contamination at the site this potentially could have given misleading results.

With shallow groundwater monitoring wells there is usually a relatively large screen particularly if the water table fluctuates and the presence of floating product is suspected.

Monitoring wells with short screen lengths should be used if possible since there is less mixing of groundwater within the well and consequently fewer tendencies to foul. If fouling is suspected then the wells need to be brushed and purged before sampling.

2.2.2 Recovery Wells

Fouling of recovery wells can be relatively rapid due to the extreme chemistry of some contaminant plumes. Contaminated water may have high concentrations of elements associated with fouling (e.g. iron, sulphide), contain high levels of organic chemical suitable for microbial utilisation and have a pH significantly different from that of the native groundwater. These problems are often increased by the groundwater mixing associated with the relatively large screened lengths required to ensure effective plume capture during pumping.

The various scenarios presented for water supply well fouling are also applicable to recovery wells. During pumping of recovery wells for shallow organic contamination the relatively reduced upper waters containing significant concentrations of Fe^{2+}, S^{2-} and occasionally Al^{3+} mix with the deeper more oxidised connate groundwater leading to fouling (Fig. 2.7). The presence of more reduced water overlying more oxidised water is the reverse of the scenario presented for redox precipitation in water supply wells. In organic rich environments where there is contact with oxygen the aerobic

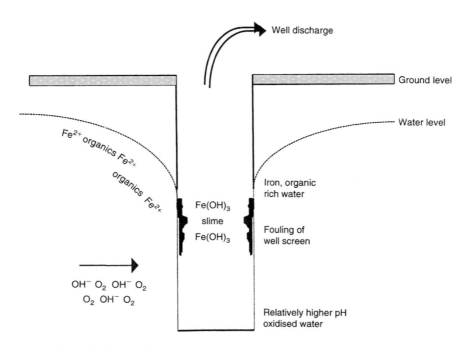

Figure 2.7 Remediation well fouling.

bacteria will create deposits with a high percentage of bacterial slime (ECP). In addition to the mixing of different types of water that occur vertically within the aquifer there can also be mixing of the waters that occur laterally around the well. If the remediation well is located on the edge of the plume then the groundwater entering the well from upgradient will be contaminated however the water entering from downgradient may be uncontaminated. At a site where a high pH, sulphide and aluminium rich contaminant plume mixed with a lower pH (iron rich) groundwater significant well fouling deposits containing carbonates and sulphides occurred in the recovery wells (Section 6.10).

2.2.3 Water Treatment Equipment

The equipment used to treat groundwater containing organic contaminants is often susceptible to impaired performance from fouling. The two most common methods for removing volatile organics from groundwater are air stripping and absorption by granulated activated carbon. In air strippers the oxidation and degassing involved in aerating the water for volatile removal can lead to oxidising any dissolved iron present causing fouling. Any carbon dioxide in the water may also be removed increasing the tendency of carbonate scales to form. In carbon adsorption units the carbon granules act as surfaces for biofilm growth that can result in severe plugging of the units. It may be possible to pre-treat the water to remove any ions susceptible to fouling or to add a biocide.

2.3 Injection Wells

Groundwater injection wells are increasingly being used in the development and rehabilitation of groundwater resources. Injection wells are used where there is limited land available for a surficial recharge basin or to inject water into deep or confined aquifers. The injection rate is commonly a third to a half of the extraction rate from an aquifer. Horizontal wells are effective in dispersing water over a wide area to a shallow aquifer. Some injection wells may also be designed to allow pumping. During groundwater remedial strategies treated groundwater is often re-injected back into the aquifer to replenish the aquifer and establish hydraulic control over the contaminant plume. Oxygen or nutrients may also be injected into an aquifer to stimulate subsurface bacterial activity. Well performance problems are commonly associated with the various applications of injection wells (Schaffner *et al.*, 1990; Payne, 1995).

The most common type of performance deterioration for injection wells is aquifer clogging. Both the physical properties of the injected water and its chemical/biological compatibility with the aquifer are important. Any suspended particles in the injected water will be filtered out within the aquifer. The suspended particle size, aquifer grain size and flow velocities will determine whether the particles are deposited close to the gravel pack–aquifer interface or further into the formation. The significant increases in yield due to backflushing these wells for short periods of time suggest that many particles are located quite close to the well. Entrapped air bubbles within the injected water will block pore spaces and reduce aquifer permeability. Oxygen dissolved in the injected water may also come out of solution if the temperature of the injected water is less than that of the aquifer. This is caused by the lower solubility of oxygen at higher water temperatures.

The dissolved organic and inorganic constituents within the injected water can react chemically with the aquifer and also stimulate bacterial activity leading to clogging. Chemical reactions can include the precipitation of various hydroxides, carbonates and sulphides. Injected waters containing high concentrations of sodium can cause the expansion of particular clays. The compatibility of the injected water and the aquifer can be assessed using chemical equilibrium software, laboratory tests and pilot scale installations (Payne, 1995). Biofouling of the aquifer can occur if the injected water contains materials that encourage biological growth. This includes nutrients, dissolved organic carbon, dissolved oxygen and dissolved iron and manganese. Large-scale waste-water re-injection schemes often use tertiary treated water to minimise the potential for biofouling. Chlorination of the injected water may also be successful. However the goal of supplementing *in situ* bioremediation is to stimulate microbial activity through the addition of nutrients or electron acceptors such as oxygen. The further these deposits are created from the well the less impact they will have on well performance in the short term although well rehabilitation is considerably more difficult. Techniques for minimising near well fouling are discussed in Section 5.4.

3 WELL MANAGEMENT

The supply of water is an essential service needed to sustain a modern lifestyle. The use of groundwater as a source of water is increasing and has some advantages over surface water. Groundwater often has a consistent water quality and is less vulnerable to pollution. But the interaction between the groundwater and a well can be complex. For example, lower groundwater water levels in an aquifer due to pumping from wells can negatively impact upon any groundwater dependent ecosystems or encourage inter-aquifer mixing potentially leading to a decline in water quality.

Although the well deterioration process is inevitable it can be managed. There is a need to choose well management strategies that make best use of both the aquifer and any groundwater extraction system so as to preserve the viable operating life of both assets. An approach to doing this is to demonstrate any costs against benefits received. This requires identifying performance indicators for the system. This structured approach to identifying the operation and maintenance needs of the well is likely to achieve the most efficient and effective operation of both the aquifer and well system.

The following terms and concepts are useful when identifying well management strategies:

- Operation
Procedures and activities involved in the abstraction of groundwater.

- Maintenance
Actions designed to keep a capital asset in a serviceable condition. The various types of maintenance activities can be defined (Janssens *et al.*, 1996):
    - Preventative maintenance involves activities aimed at interruption of the deterioration process at an early stage based on systematic, prescheduled programmes of providing early detection of beginning effects. Monitoring is the regular observation and recording of information from wells and forms the basis of preventative maintenance.
    - Corrective maintenance is activities in response to breakdown or detected defects. It restores the facility to good working condition without significantly expanding it beyond its originally designed function or extent. Rehabilitation is an extreme form of maintenance due to the degree of action required. It is carried out to correct major defects, wear and tear, often as a result of inadequate maintenance.
    - Design-out maintenance focuses on improvements to equipment design to simplify maintenance operations and/or increase reliability.

3.1 Well Maintenance Strategies

An efficient and effective well maintenance strategy involves finding the right mix of preventive and corrective maintenance activities. It can be difficult to judge this in environments where equipment failure behaviour is poorly known and the cost of various maintenance activities to achieve a given level of well performance is difficult to estimate. This has led to a diverse range of strategies to evaluate and manage well maintenance programmes. The nature of maintenance work undertaken and the standard to which this is done will often reflect local judgement, customs and practices, available resources and the type of equipment within the well system.

Several types of strategies can be identified:

- Failure based strategy
This strategy involves repairing equipment only after failure. There are minimal costs associated with monitoring well performance however the rehabilitation costs can be high. This is a high-risk strategy that can lead to poor system reliability.

- Prevention based strategy
This involves routine replacement work based on the estimated service life of equipment. This strategy will be effective only when the well system is well characterised and there are reliable statistics on equipment failure. In other environments it can lead to either a cost intensive approach or create system unreliability due to equipment failure.

- Performance based strategy
Activities occur when a performance indicator for the system fails to reach a predetermined standard. Assessment of whether these performance indicators are being met is dependent upon suitable information being collected during a monitoring programme. The performance indicators allow the early detection of problems that could cause system unreliability through equipment failure or rehabilitation work.

3.2 System Performance Indicators

The following are some factors to consider when developing performance indicators for a well or well system. The type of well and its purpose will determine which of these indicators are relevant for a particular well. The acceptable range of these indicators should also be specified.

- Hydraulic
This can be represented by instantaneous flow rates or daily volume of water pumped. A threshold for triggering rehabilitation due to decreased well efficiency is often stated as 60–75% of the original performance.

• Water quality
There are water quality guidelines which relate to public health, aesthetic and equipment maintenance criteria. Concentrations of greater than 0.3 mg/L iron and 0.05 mg/L manganese can stain plumbing, fitting or washing. A practical limit for sand content on pump wear is often 5–8 mg/L.

• Structural
This involves the structural integrity of the system as well as wear and tear on the pumping system.

• Economic
Power consumption is a key economic performance indicator. This can be measured as the power costs for a given volume of water pumped (wire to water ratio).

• Maintenance/monitoring costs
This can be evaluated from the costs of monitoring activities and corrective maintenance.

• Social
The social factors are difficult to quantify but reflect user expectations for the delivery of a safe and reliable water supply. This can vary depending on the end use (domestic, agricultural and industrial), size of the population served and reliance on the water supply (domestic/urban, rural/agricultural). The availability of alternate water supplies will impact these factors.

• Environmental
The impact of the system upon the environment may also be important and constitute a performance indicator. The lowering of the regional water table can impact on other users and vegetation. Also noise and aquifer contamination from well operation may need to be considered.

3.3 Evaluating Well Maintenance Components

The development of a well maintenance strategy involves identifying what the indicators of adequate well performance are and then developing a suite of monitoring and maintenance activities to ensure these are met.

For well construction there have been minimum design standards proposed which can be used to standardise costs. It is not possible to develop a costing index for well maintenance activities. The iterative problem solving nature of the work and site specific variables make it difficult to assess the extent of the work required. Some maintenance activities can be carried out in-house while other services may need to be contracted out. There can be considerable diversity in the availability of contractors who have the knowledge and equipment to perform the required work.

The following sections identify both a list of actions that can be involved in these activities and a generic list of the components to be considered in costing these activities.

3.3.1 Costing Monitoring and Maintenance Activities

Costing of these activities may involve both services from the staff that operate the wells and external contractors. The following is a list of possible costing categories (ODA, 1993a–c):
- Equipment: purchase, depreciation, insurance, safety measures/equipment, licensing
- Labour: unit costs, insurance and overheads, training and support, accommodation, overtime and allowances
- Transport: vehicle purchase, depreciation, insurance, running costs for personnel/ equipment/sample transport
- Energy: fuel, electricity
- Other: contracted services (e.g. water analysis)

3.3.2 Monitoring Costs

A range of activities can be involved in monitoring well performance (ODA, 1993a–c):
- Keep and analyse operational records
- Evaluate pump performance
- Evaluate well performance
- Well logging
- Water physical tests
- Water chemical tests
- Water bacterial tests
- Visual inspection of headworks
- Pump recovery and inspection

3.3.3 Maintenance Costs

The maintenance requirements of a groundwater extraction system can vary significantly between its components. The pumps and prime movers often have a higher level of maintenance required compared with the well casing.

Well casing may require the following mechanical treatments:
- Brushing
- Surging/swabbing
- Jetting
- Solids removal

These treatments are often combined with various chemical treatments:
- Oxidants
- Acids
- Surfactants

A typical well cleaning operation could include:
1. Measurement of the specific capacity, water pH and water electrical conductivity (EC)
2. Removal of the pump from well
3. Brushing of the well casing and screen
4. Removal of wastewater either by pumping or bailing
5. Shock chlorination
6. Repeat of step 4
7. Acid treatment of well using surging or jetting
8. Repeat of step 4
9. Optional; repeat steps 5–8
10. Replacement of pump
11. Measurement of the specific capacity, ensuring the water is the original pH ± 0.5 units and the EC is within 10% of the original value

An important consideration can be the treatment and disposal of the wastewater generated during these treatments. The time and cost associated with removing wastewater from the well can be minimised if the contractor has a pumping plant designed for rapid insertion and removal from the well.

The maintenance of the pumping system can involve:
- Lubrication
- Cleaning of parts
- Repair and replacement of parts

The frequency with which these activities need to be undertaken is dependent on many factors including aquifer type, well design and operating procedures. Some published data on the frequency of these activities for municipal wells has been included to assist in the planning of long-term maintenance requirements where there is not sufficient information to establish this from wells within the local area (Table 3.1).

3.3.4 Replacement Well Costs

Installing a replacement well involves decommissioning or modifying the existing well and constructing a new well. The well being abandoned may still retain some financial value if it is used as a monitoring well.

Table 3.1 Types of well problems in different environments.

Aquifer type	Most prevalent well problems*	Major maintenance frequency** (years)
Alluvium	Silt, clay, sand production, iron and manganese biofouling, limited recharge, casing failure	2–5
Sandstone	Fissure plugging, sand production, casing failure, corrosion	6–8
Limestone	Fissure plugging, carbonate fouling	5–12
Basaltic lavas	Fissure plugging, fouling	6–12
Metamorphic	Fissure plugging	12–15
Interbedded sandstone and shale	Drilling mud invasion during drilling, fissure plugging, limited recharge, casing failure	4–7
Consolidated sedimentary	Fissure plugging, biofouling, fouling	6–8

* Excludes pumps and declining water tables. (Modified Gass *et al.*, 1980.)
** Based on the following assumptions: Wells are pumped continuously at their highest sustained yield; major maintenance is required when the sustained yield is less than 60% of the initial yield or the well efficiency is less than 60%; major maintenance represents a cost expenditure of approximately 10% of the current replacement cost. Minor maintenance is excluded; wells are designed in accordance with current practices, not necessarily in accordance with best available technology.

The activities involved in decommissioning can include:
- Removing the pump and headworks
- Sealing/grouting the well
- Capping of the well
- Getting regulatory approval

The construction of a new well can involve:
- Regulatory approval and license
- Site survey
- Well drilling and construction
- Pumping equipment
- Headworks and other infrastructure

4 DIAGNOSING WELL DETERIORATION

The performance of a groundwater pumping system is dependent upon the efficiency of the well, aquifer and pump. It is necessary to distinguish between these components when diagnosing well deterioration problems.

4.1 Diagnosing Well Deterioration

There are two types of evidence that may trigger a detailed examination of the performance of a well. These are a decrease in the water quality and/or the reduced yield of a well. Some problems may only impact on one well performance criteria. Corrosion of the casing is likely to only impact upon water quality by allowing sand or water entry. However with other types of problems both performance criteria associated with water quality and water yield may be affected. With iron biofouling there will be a decline in water quality and possibly a decline in water yield. An accurate diagnosis of the problem will ensure that a greater range of options for controlling the deterioration process is available.

4.1.1 Water Yield Decrease

To establish that there has been a decline in the capacity of the well system to produce water requires the analysis of any well monitoring data. It is important to have a rigorous approach to determining the well deterioration causes and actions needed (Fig. 4.1). A decline in the wire to water ratio with time demonstrates that the well and/or the pump are/is hydraulically impaired. It is important to ensure that system characteristics (e.g. the total pumping head) are accurately known when comparing the data. Distinguishing between well and pump deterioration requires the analysis of the pump performance data. If there has been a decline in the performance of the pump then there may be fouling deposits in the pump–riser pipes or wear on the pump impellers. If the pump performance is adequate then either the well, aquifer or pipeline have reduced the well yield. A review of the aquifer monitoring data can establish if the available drawdown has increased due to a lowering of the regional water table. Changes in the condition of any pipeline used to transmit water from the wellhead can also be assessed. Specific capacity data from the well will indicate a change in the efficiency of the well. A step drawdown test will be needed to determine whether this is due to blocking of the well screen or aquifer. The composition and nature of the deposit can be determined from water and solid sample analysis. Well inspection techniques can be used to determine the location, nature and extent of the problem.

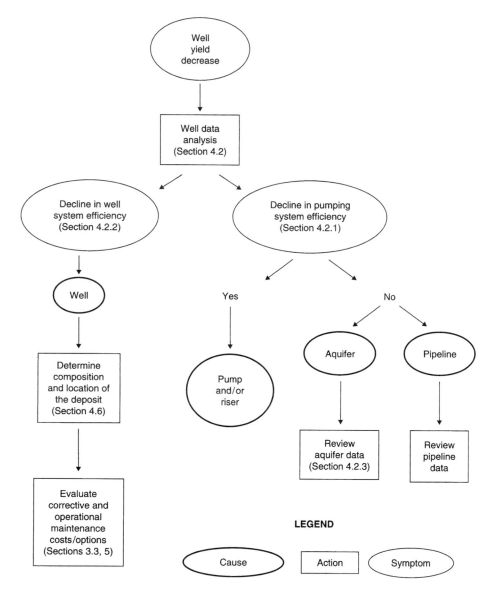

Figure 4.1 Diagnosing water well yield losses.

4.1.2 Water Quality Decrease

A decrease in the quality of the water discharged from the pump can indicate the well is in need of corrective maintenance (Fig. 4.2). Both water and solid material can be sampled (Appendix E) to identify nature of the problem (Sections 4.3–4.5). The water quality can be used to infer the nature of the problem, which can then be confirmed by examination of any solid material present. The most common water quality parameters,

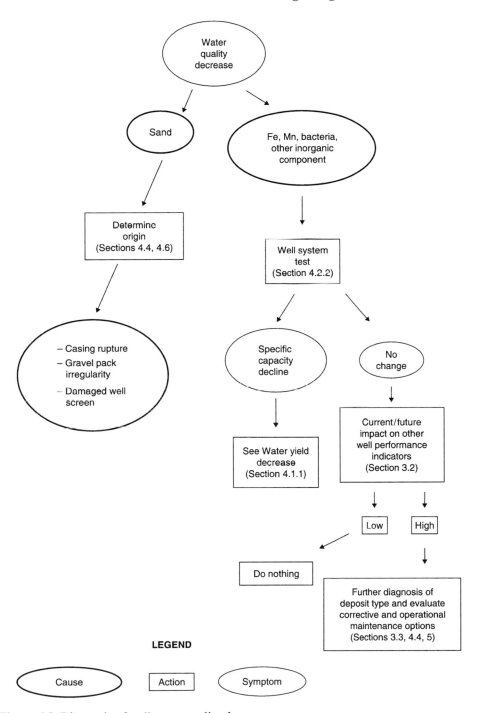

Figure 4.2 Diagnosis of well water quality decreases.

which can change due to well deterioration, are sand content, iron and manganese levels and bacterial counts.

If sand is present then a visual analysis using grain size and type will indicate the origin of the particles. The cause of the problem may be rupture of the casing, an irregularity or bridge in the gravel pack or damage to the well screen. The location of the entry point can be localised by using well inspection (Section 4.6) or by surging the well. During surging a bailer is lowered to the bottom of the well and the well is surged in the section of casing immediately above it. The bailer is retrieved and inspected for sand and then emptied. The procedure is carried out progressively up the well until no more sand occurs in the bailer. The last section where sand occurred in the bailer contains the ruptured casing.

If elevated iron, manganese or bacterial levels are measured then biofouling may exist in the well. The hydraulic performance should then be established through examining the operational records from the well and conducting specific capacity tests (Section 4.2). If there is a decline in the specific capacity then biofouling deposits are hydraulically impairing the well and correctional maintenance may be needed. If there is no change in the hydraulic performance of the well then the significance and implications of the water quality change needs to be assessed in terms of the other well performance indicators (Section 3.2).

4.2 Well Data Analysis

The total performance of a groundwater extraction system is dependent upon the well, aquifer and the pumping system. It is necessary to distinguish the contribution of each of these components when diagnosing well problems. The data required to assess system performance can be obtained by monitoring during the normal operation of the system or from specialised tests designed to assess the performance of individual components. Particularly when assessing operating data collected over a long period it is necessary to check the reliability of the data (Section 4.2.4). Since fouling problems occur over a number of years, good field observations can help considerably in understanding the problem.

4.2.1 Pumping System

The overall pumping plant efficiency is influenced by the:
• Pump
• Prime mover (e.g. electric motor, internal combustion engine)
• Drive system: transmission between the prime mover and pump (e.g. direct coupled, flat belt, vee belt, gears)
• Hydraulic system from the pump intake to the discharge (e.g. pipes, shafts, tubes)
• Electrical system

Each of these components in the pumping plant will have different efficiencies for converting input energy to output energy.

Pumping Plant Efficiency: Wire to Water Ratio
A measure of the overall pumping plant efficiency is the wire to water ratio. This ratio represents the efficiency of the pumping system that converts input horsepower (IHP) to water horsepower (WHP):

$$WW = \frac{WHP}{IHP} = \frac{Q * TDH * 0.65}{3956 * kW} \qquad (4.1)$$

where

WW = wire to water ratio
WHP = energy required to pump water against a given head
IHP = energy consumed by the prime mover
Q = discharge rate (L/min)
kW = power input (kW)
TDH = total dynamic head (m)

The WHP can be calculated from the discharge rate and TDH. The TDH includes all head losses in pipes and the head difference between the pumping water level and the free flowing discharge point. Losses in pipes and fittings can be obtained from an Engineering Handbook. The IHP can be calculated for electric motors from the watt-hour meter installed on the control panel. The IHP for internal combustion engines can be calculated from methods described in an Engineering Handbook or pump manufacturer literature.

Pump performance curves show the relationship between the head capacity, brake horsepower (power required by the pump), pump efficiency and discharge capacity. A pump performance curve can be obtained from the pump manufacturer or generated from data obtained during a step drawdown test. The efficiencies measured from an installed pumping plant will be slightly lower than that in the pump manufacturer literature. The measured pumping plant efficiency incorporates loss due to both the pump, prime mover and drive system while the manufacturers' literature only includes pump efficiency losses. Typically an electric motor is 90–95% efficient.

The data collected for a pumping plant efficiency test can be used and presented in several ways:
1. Deterioration/Suitability of the system. This can be established by calculating the measured pumping plant efficiencies (wire to water ratio). Typical pumping plant efficiencies are around 65–75% (Driscoll, 1986; Helweg *et al.*, 1983). If the values fall below this the system may require maintenance or be mismatched to the current hydraulic conditions in the well. A decline can indicate wear on the pump components, changes in friction losses in pipes due to fouling or aquifer water level changes.

2. Cost of the water. The kilowatt-hours consumed per megalitre of water produced can be calculated. This can then be converted to a cost per megalitre. This allows an economic evaluation of operating costs for different wells to be undertaken.

Pumping plant efficiencies and specific capacities are not constant across a range of discharge rates. The data used to calculate the pumping plant efficiency can be collected from measurements taken during normal pump operation or from step drawdown tests. A step drawdown test requires well recovery to its static or pre-pumping water level. The pumping rate is then increased incrementally for at least three steps while the drawdown is measured. A typical range of flow rates would be 25%, 50%, 75%, 100% and 125% of the well's normal operating discharge rate. The procedure for conducting and analysing a step drawdown test is discussed elsewhere (Kruseman & de Ridder, 1990; Driscoll, 1986; Clarke, 1977) and in many hydrogeology textbooks. A good example of the use of step drawdown tests to evaluate well deterioration and well treatment methods is described by Clarke and others (1988). It is also possible to acquire specific capacity data during the normal operation of the pump. However because the operating conditions can vary considerably between measurements the data is considered less reliable than that from step drawdown tests.

Consideration should also be given to the reliability of the pumping system equipment (Section 4.2.4) when interpreting data collected over long time periods in fouling environments. In a wellfield the TDH needs to be carefully monitored since the pipeline pressure can vary significantly depending on the number of pumps in use at the time of measurement.

4.2.2 Well System

To evaluate the performance of a well the amount of drawdown for a specified discharge rate must be established. The amount of drawdown represents the energy losses in both the aquifer and the well. It can therefore be used to measure changes in the efficiency of the well system:

$$s = BQ + CQ^n \qquad (4.2)$$

where

> s = drawdown
> Q = discharge rate
> B = aquifer loss coefficient
> C = well loss coefficient
> n = constant (often approximated by 2)

The aquifer loss (B) generally occurs when the water passes through the aquifer to the well. The flow is often laminar and so is linearly proportional to the discharge from the well. The well loss (C) occurs as water enters the well through the well screen. Turbulent flow occurs in this zone due to the higher velocity of the water. The energy losses are much greater and are exponentially related to the well discharge. The greater the value of C the greater losses attributable to well loss. This can be due to clogging of the well screen slots. The terms aquifer loss and well loss while conceptually convenient may not be accurate in all situations. The aquifer loss term (BQ) can include some elements of losses due to water entering the well (Clarke *et al.*, 1988) under laminar flow.

The performance of a well can be analysed using specific capacities, well efficiency or through determining aquifer and well losses.

4.2.2.1 Aquifer and Well Losses

The most comprehensive analysis of the performance of the well and aquifer is by determining the aquifer and well losses that are represented by B and C in equation 4.2. This analysis may allow the identification of clogging that has occurred at the interface between the aquifer and the gravel pack or in the formation to be distinguished from that occurring at the well screen. This information can be important in the design and choice of treatment during well rehabilitation (Clarke *et al.*, 1988). The procedure uses the specific capacity data obtained during a step draw-down test.

4.2.2.2 Specific Capacity

The specific capacity of a well is a measure of the productivity of the well. It includes losses attributable to both the aquifer and the well but not the pump. This analysis will give information on whether there has been a change in ability of the well to produce water, however it cannot distinguish whether this is due to a change in the condition of the well or the aquifer. That can only be evaluated from the well and aquifer losses. The specific capacity can be calculated from:

$$SC = \frac{Q}{s}$$
(4.3)

where

        SC = specific capacity
        Q = discharge rate
        s = drawdown

The specific capacity can be measured during a step drawdown test or monitored during normal operating conditions. A step drawdown test can be undertaken with the discharge rates selected to replicate the original test conditions. This considerably simplifies the analysis since the current data can be directly compared with the original data when the system was new. Any changes can then be attributed to system deterioration. If the discharge rates are different from those measured during the original tests then the data needs to be normalised. This involves calculating what the original pumping plant efficiency or specific capacity would have been for that discharge rate. This can be estimated by plotting the original data across a range of discharge rates and then estimating a value from the curve for the discharge rate used in the current test.

An advantage of a step drawdown test is that specific conditions are required for the test procedure, which leads to greater data reliability. Both the pre-pumping water level and the time of pumping when drawdown measurement are taken are accurately known. The specific capacity data collected during normal pump operation is often subject to some variability, but the savings in cost and manpower over the more comprehensive step drawdown test can be significant.

The measurements required for the determination of specific capacity during normal pump operation are:

• Discharge rate

• Pumping water level

• Static water level
If the water table is reasonably stable through time then the static water level can be determined from the water level when the well or nearby wells were recently shut down. It can also be monitored from observation wells outside the cone of influence or those showing minimal drawdown.

• Time of pumping
It is important to know how long the pump had been operating when the pumping water level was taken. Wells may take from a dozen hours to several days (dependent upon aquifer type) to reach a steady state drawdown. Measurements taken during normal pump operation assume a steady state drawdown, which may be significantly different to that measured after the set pumping period allocated during the step drawdown test (often 1 h). Knowing this information will allow a comparison with data from step drawdown tests and data collected at other times during well operation. When comparing data collected under different prior pumping conditions it is important to identify the sensitivity of the data to errors that could have occurred in its measurement. For example, the biofouling of the flowmeter could cause errors of 20%, what effect would this have on the calculated specific capacity? If the well takes one day to reach a steady state pumping level, what is the difference in drawdown if it is measured after 1 h?

Analysis of flow rate and drawdown data on the well allows the specific capacity of the well to be plotted with time to determine whether there actually has been a decline and whether a decline is sudden or gradual. This approach is adequate as long as conditions within the aquifer system have remained constant. In a wellfield subject to biofouling it is good practice to have annual specific capacity pumping tests with intermittent data provided by specific capacities calculated during normal pump operation.

4.2.2.3 Well Efficiency
The efficiency of a well can be calculated by dividing the existing specific capacity by the maximum specific capacity measured for the well:

$$WE = \frac{SC_e}{SC_m} \qquad (4.4)$$

where

    WE = well efficiency
    SC_e = existing specific capacity
    SC_m = maximum specific capacity

The specific capacity (maximum) value is often the specific capacity measured when the well was new. However, the specific capacity of some wells in sedimentary aquifers can improve after construction due to development of the aquifer during extended pumping. It is the highest consistent value that should be used as the maximum specific capacity.

A well that has declined to an efficiency of 60–80% is often considered to be in need of corrective maintenance. It may be significantly more difficult to return a well to its original performance when the well efficiency has dropped to a low level. The exact well efficiency at which it is prudent to rehabilitate the well depends on what the important performance indicators are for a particular well (Section 3.2) and the cause of well deterioration. Having reliable and detailed data on the decline in well performance allows the well manager to confidently predict the best time to rehabilitate a well.

4.2.3 Aquifer System

The ability of a well to keep on producing the discharge rate obtained during well commissioning is reliant on the aquifer conditions remaining stable. A decline in the available well drawdown will cause a decline in well yield.

The available drawdown can be reduced by the
- Additional drawdown from nearby wells, which have been newly established, or their discharge rate increased
- Decrease in groundwater recharge
- Impact of a groundwater flow boundary intersected as the cone of drawdown has increased during pumping
- Changes in aquifer permeability

4.2.4 Data Reliability

Significant funds can be invested in the acquisition, operation, and maintenance of water wells. It is important to ensure that the data upon which decisions are made regarding the management of the well are reliable and accurate. In a wellfield where fouling is suspected it is important to validate the accuracy of the data to be used in assessing either pump or well efficiency.

When wells are connected into a pipeline the pressure in the pipeline will impact upon the hydraulic performance of the pump. The headloss in a pipe is impacted by the flow diameter and flow surface characteristics (e.g. surface roughness). These parameters can change due to the buildup of fouling deposits leading to an increase in headloss. However fouling of the riser pipe from the pump can only be detected by inspection or by comprehensive data analysis of the pump performance rather than the well performance. This type of fouling will increase the TDH causing a decline in the wire to water ratio, which if undetected could be attributed to other causes.

Many types of discharge flowmeters are susceptible to fouling. It is common for the impellers on the spindle type flowmeters to build up a deposit, which will increase the velocity of water within the meter causing an increase in the measured discharge rate. This causes an apparent increase in discharge, which increases the wire to water ratio and the specific capacity even when there may be a decline in system performance. The data collected prior and post cleaning of the pump equipment should be carefully evaluated for abrupt changes, which indicate there may be problems with data reliability of the pump unit or associated equipment.

Wells subject to performance deterioration may require frequent pump removal. Errors in the drawdown measured by air-lines can occur after pump removal, unless care is taken to attach the air-line back to the same place on the riser pipe.

4.3 Water Sample Analysis

4.3.1 Water Quality Indicators

Analysis of water samples can give an indication of the processes occurring within a pumping well. The presence or a change in the concentration of specific water quality indicators can be used to monitor fouling and corrosion processes (Table 4.1). The accuracy of this approach depends on the fouling/corrosion process producing byproducts that can be removed by the pumped water from where the activity is taking place and then be measured at the wellhead. Since biofouling generally produces more soft removable components than corrosion or carbonate fouling it is this process that is most likely to be detected based on water quality indicators.

The most important water quality indicator is the level of iron in the water. This will indicate the presence of iron biofouling, which is the most common type of well

Table 4.1 Water quality indicators of well deterioration.

Process	Water quality indicator	Response
Biofouling	Fe^{2+}, Fe(total), turbidity	Erratic
	Sulphide, H_2S	Increase
	Bacterial numbers	Increase
Corrosion (steel)	Turbidity, suspended solids, Fe^{2+}	Increase

deterioration. Sulphide production, which can occur within a biofilm, may also be a useful indicator. Samples collected for determination of Fe^{2+}, sulphide and Eh can change significantly if exposed to air and careful sampling techniques are required (Appendix E). It is suggested that Fe(total) which includes the concentrations of both Fe^{2+} and Fe^{3+} should always be measured as this will include iron that was oxidised during sampling as well as colloidal iron. Very localised corrosion pitting is difficult to detect while more uniform corrosion across a larger surface will produce more corrosion byproducts and be detected. If the well corrosion results in the entry of aquifer material or different water types then a change in the water chemistry can be detected at the wellhead. Significant changes in general water quality parameters (e.g. pH, Eh, EC, major cations/anions) can indicate corrosion of the well casing has allowed different types of water into the well. The extent to which corrosion actually occurs cannot be predicted or diagnosed by enumerating specific groups of bacteria (Little *et al.*, 2000).

4.3.2 Fouling/Corrosion Predictors

There have been many attempts to relate either fouling or corrosion to water quality but no indicators have been found to be universally applicable. Evaluation of fouling/corrosion predictors can give an idea of the dominant processes, which may be causing or likely to cause fouling, or corrosion. These predictors only take into account the inorganic precipitation of compounds and do not include microbial factors, the interaction with other compounds in solution (e.g. chelates) or the rate at which the reaction will occur. The significance of these factors which are not included and where the water samples selected for analysis were collected will determine the accuracy of the predictor method.

The simplest method of estimating the tendency of a dissolved species in groundwater to precipitate is through the use of an Eh–pH or pH–solubility diagram. The equilibrium solubility of a dissolved species is related to the pH and in some cases to the Eh of the groundwater. By measuring the pH and Eh of the water the likelihood of that species precipitating under those conditions can be estimated. The diagrams for the chemical species involved in well fouling are presented in Appendix A. The mass of precipitate that will form is then dependent upon the concentration of the dissolved species. This approach is useful when evaluating the impact of mixing waters of differing hydrochemistry either within a well from different aquifers or during water injection. The diagrams are based on very simple water compositions and more complex situations will require the use of a more sophisticated approach such as geochemical computer programs or laboratory tests (Payne, 1995). The tendency of water to form a calcium carbonate scale can be estimated from Langelier Saturation Index (LSI) and Ryznars Indexes (RI).

The degree of metal corrosion will be influenced by the chemical and physical properties of the groundwater as well as metal composition. The relevant groundwater properties are salinity, dissolved oxygen, and carbon dioxide, flow rate, pH and temperature. However predicting whether corrosion will occur for particular water chemistry is difficult. The standard corrosion predictors (e.g. LSI, RI) relate the tendency of water to produce a carbonate scale with the non-corrosivity of the water. Therefore when the conditions are suitable for the precipitation of a calcium carbonate scale then this will protect the metal surface from corrosion. These indices through calculating the tendency of a carbonate scale to occur will only imply the likelihood of corrosion rather than the degree of corrosion. It is generally recognised that the LSI may only indicate the corrosive tendency of water within a pH range 6.5–9.5 (AWWA, 1986). An LSI of greater than zero indicates that the formation of a carbonate film is likely. An LSI of equal to zero or a negative value indicates scale formation is unlikely. Another approach is to use corrosion data that is published in the literature based on field studies of groundwater wells. The data from a research study is presented in Appendix D. The water chemistry in a well of interest can be compared with similar wells used in the corrosion study and the corresponding corrosion rate used. This should only be considered a screening level approach since the operating schedule of the well and differences in the water chemistry will affect the corrosion rate. The only way to confidently estimate the corrosion rate would be to conduct a corrosion test using test coupons (Appendix D) on a well in the same aquifer. This more sophisticated approach is more suited to the development of a wellfield rather than construction of a single well.

4.4 Solid Sample Analysis

An understanding of the composition of a fouling deposit is an integral part of unravelling the processes within the well and aquifer that have lead to the formation of the deposit. The chemical species present can allow an inference of the process that has caused them to precipitate. The ratio of inorganic to organic matter can indicate the significance of microbes in the fouling and may affect the choice of treatment chemicals. In some cases it is useful to know the types of bacteria present.

The resources devoted to the sample analysis program depend on the extent of the problem, analytical facilities available and the need to solve the problem. A range of analysis from field to laboratory based techniques can be used in examining solid samples. Procedures for obtaining solid samples are detailed in Appendix E2.

4.4.1 Visual

Some preliminary ideas on soft deposits may be gauged from the colour of the pumped water. If the water is red colour then iron hydroxide deposits will be present while 'black' water indicates iron sulphide or manganese. Fouling deposits can have a large range in colour but the presence of iron will tend to tint the deposit towards a red colour. This occurs even in carbonate deposits that are normally white and hard.

The type and size of any sand in the pumped water or in a well deposit can be used to infer its origin. If the sand is smaller than the well screen slots then either the well is still being developed or the gravel pack may be bridged and the formation is in contact with the screen. If the sand is larger than the slot size then the casing may have ruptured or the slot size may have been enlarged by corrosion. If there is gravel pack in the sample then the casing has ruptured.

4.4.2 Odour

An organic odour can indicate the origin of some deposits. Biofouling deposits often contain protein that smells like burnt hair upon heating. The odour of rotten eggs indicates sulphides are present.

4.4.3 Acid Treatment

Calcium carbonate vigorously bubbles upon contact with an acid while calcium–magnesium carbonate has a slower reaction. Sulphate compounds do not bubble in contact with acid however iron sulphide will react and form hydrogen sulphide (rotten egg gas).

4.4.4 Loss on Ignition

The loss on ignition (LOI) method involves heating a solid sample, which causes specific compounds to form gases and vaporise. At selected temperatures specific reactions will occur within the sample and the measured weight loss is related to the amount of the compound initially present. This technique is useful for evaluating the ratio of organic to inorganic matter:

- 105°C
Heat a sample in an open container for at least 1 h, cool in a desiccator and then weigh. The difference in weight before and after heating represents water trapped in the deposit.

- 550°C
Heat a sample (previously heated to 105°C) for at least 1 h at 550°C. The difference in weight represents the volatile organic fraction. The temperature used to establish the transition between the organic and inorganic fractions (above 550°C) is somewhat flexible with Standard Methods (APHA, AWWA & WCPF, 1992) using 550 ± 50°C. Caution should be maintained if the sample is derived from a contaminated site since the off-gas may be hazardous.

- 1050°C
Heat a sample (previously heated to 550°C) for at least 1 h at 1050°C. The difference in weight between 550°C and 1050°C represents carbonates and sulphur. The weight

Table 4.2 Typical LOI data for well deposits.

LOI400°C (% weight)	LOI1050°C (% weight)	Average ratio* (%)	Description of deposit
1	44	2	White, calcium carbonate
1–5	32–43	8	Red-black, calcium carbonate with some iron present
10–18	15–50	48–85	Red, biofouling deposit with iron present
21–60	42–68	83	Yellow-brown, biofouling deposit with aluminium, sulphur present

* LOI400°C divided by LOI1050°C.

of the remaining sample represents all the other chemical species present, which can then be identified by X-ray fluorescence (XRF).

An analysis of deposits from 44 wells found that the LOI data varied between different types of deposits (McLaughlan *et al.*, 1993). The ratio of LOI400°C to LOI1050°C clearly shows the deposits that have a large organic component. In these samples microbiological activity has played a significant role in the formation of the deposit and the selection of chemicals for well treatment should take this into account (Table 4.2).

4.4.5 Total Organic Carbon

The total carbon in a sample is comprised of both inorganic and organic carbon. A significant organic carbon concentration indicates biofouling or contamination with organic chemicals. The total organic carbon (TOC) present in a water supply well deposit gives an indication of the amount of biologically derived material within the deposit. The analysis methods are outlined in Standard Methods (APHA, AWWA & WPCF, 1992).

4.4.6 X-Ray Fluorescence

This is a convenient and relatively cheap method used extensively by geochemists to get a broad spectrum of elements. This technique has minimal sample preparation, which is an advantage over some other methods that require the samples to be acid digested. The elements present in the sample can indicate the type of deposit and what caused it to form (Table 4.3). A full classification of well fouling deposits is presented in Appendix B.

4.4.7 X-Ray Diffraction

This technique can be useful to determine crystalline compounds such as carbonates and clay occur within the deposit. Many of the iron containing compounds precipitate

Table 4.3 Interpretation of elements present in a deposit.

Element	Comments
Si	Sand, silt or clay present
Ca, Mg	Carbonates present
Al	In significant amounts it may indicate aluminium fouling while in small amounts it will indicate clay is present
Fe, Mn	If this is the dominant element then the deposit probably has a microbial component

very quickly and lack a well-defined crystal structure that can make them difficult to identify by X-ray diffraction (XRD).

4.5 Microbial Sample Analysis

It is now recognised that most groundwater environments support a broad microbiological population. The need to substantiate through microbial analysis that microbes are present in the groundwater or aquifer material depends on the type of fouling and the level of investigation to be undertaken. Microbial analysis of deposits may be considered both as having both a confirmation and a process definition role. For iron fouling deposits it is widely recognised that bacteria will exist and subsequent analysis of the deposits should be considered as confirmation. Their involvement in the reactions occurring in groundwater ensures their association with iron fouling deposits. However to establish whether there is direct involvement by manganese oxidising or sulphate reducing bacteria would require confirmation from microbial analysis. The extent to which corrosion actually occurs cannot be predicted or diagnosed by enumerating specific groups of bacteria (Little *et al.*, 2000). The microbes associated with biofouling may comprise both cultivable (e.g. heterotrophs) and non-cultivable (e.g. filamentous bacteria) types.

4.5.1 Examination

The direct examination of water and solid samples by light microscope can show fragments of the stalked and filamentous bacteria that are currently not commonly cultivable. However bacterial structures from cultivable bacteria may also be apparent. There are various stains available which further identify the bacteria and their byproducts. The commonly used methods are documented in Smith (1995) and the Standards (APHA, AWWA & WPCF, 1992) and ASTM D932-85. A light microscope with a magnification of 200 times to 1000 times is suitable.

4.5.2 Culturing

The types of bacteria isolated from water or solid samples are dependent upon the media used. The media serves as an enrichment tool to enhance the growth of selected microbe types, but how well the media cultures the active biofouling populations are unknown. There are both field and laboratory based methods available.

4.5.2.1 BART

The BART method involves the addition of a water sample to a tube filled with a specialised dehydrated media (Cullimore, 1992) and can be carried out in the field. The response based on the time before a noticeable reaction occurs and the appearance of the resulting reactions can give qualitative results (Smith, 1995). These kits are available commercially (e.g. Hach Chemicals) for iron related bacteria, sulphate reducing bacteria and slime forming bacteria. The iron related kit cultivates the microaerobic heterotrophs only.

4.5.2.2 Standard Methods

There are widely recognised laboratory methods for culturing iron precipitating, manganese oxidising bacteria as well and *Gallionella* enrichment presented in Standard methods (APHA, AWWA & WPCF, 1992). Specialised formulations also exist in the research literature (Smith, 1995).

4.5.3 Collection

Groundwater microbes can be collected from the well either in a water sample or as part of a solid sample from a biofouling deposit. In collecting a water sample it is important to realise that the bacteria present in that sample may be free floating bacteria that do not contribute directly to the fouling process. They may also be sessile bacteria that do contribute to the process but have been washed away from a biofilm during pumping. Therefore a water sample may not contain the microbial diversity of a solid sample (either aquifer material or biofouling deposit). But it is easier to collect water samples than fouling deposits. In collecting water and solid samples the guidance outlined in Appendix E should be considered.

Solid samples can be collected directly from biofouled surfaces. Where there is no direct access to fouled surfaces the solids samples may be filtered/cultured on prepared surfaces placed in contact with the groundwater. Sampling devices can either be lowered down the well under the water level or connected up to the wellhead. The time required to collect an adequate quantity of sample by these devices depends on the degree of filtering involved in the method chosen. The filter sampling method may require 1 h while the coupon devices that rely solely on bacterial growth rather than filtering could be in excess of one week. These remote-sampling methods can also be used in a monitoring regime to determine whether biofouling has occurred or the conditions are suitable for it to occur.

4.5.3.1 Filters

Membrane filters (0.45 μm) can be connected at the wellhead to collect suspended sediments from the groundwater. The sediment-laden water obtained immediately after pump startup of a biofouled well will contain material dislodged from deposits with the well and riser pipe. This method can also be used for to assess the clogging potential of particles in water for subsurface re-injection. The method is described in NACE Test Method (1984) and Schippers *et al.* (1995).

4.5.3.2 Moncell

This device connected at the wellhead consists of a tube packed with a porous media (sand, glass beads) through which groundwater is passed (Howsam & Tyrell, 1989). The material collected comprises both suspended particles filtered from the groundwater and any deposits formed on the porous media while the device is installed.

4.5.3.3 Coupon Devices

Coupons can either be placed within the well if access permits or be connected to a water flow bypass at the wellhead. These devices do not filter the water, but instead rely upon growth upon the immersed surface. The inwell devices may consist of a series of slides (coupons) within a holder lowered into the well (Smith, 1992) while at the wellhead a watertight coupon holder is used. The growth conditions for the inwell sampler more closely approximate those occurring on well surfaces. The *in situ* collector needs to be retrieved and placed. There can be problems with access ports at the wellhead, clearance between the riser main or pump and with the well casing and entanglement of the device around electrical cables.

From a practical point of view the composition of the surface (e.g. steel, plastic, glass) has little influence on the microbiology of the deposit. Glass slides would be preferred for direct microscopic inspection. An alternate method for collecting biomass is to lower a weighted rope into the well and allow biofouling deposits to buildup on it. This method has the advantage of determining where vertically within the well the buildup of fouling is greatest, but obtaining a sample for microscopic examination requires disruption of the biofilm.

4.6 Well Inspection

The condition of the well casing can be determined visually with a CCTV camera or through geophysical logging. The visual appearance of casing from a CCTV log may show the structural condition of the inside of the casing wall but gives no indication of the external condition. Both CCTV and geophysical logs are complementary and useful tools in well diagnosis (Bliss, 1990). The type of logs used will depend on the equipment available, budget and type of problem.

The types of geophysical logs useful for well rehabilitation investigations include:
- Caliper
- Density
- Temperature and conductivity
- Sonic
- CCTV

A caliper tool has three or four arms that are expanded to rest against the inside of the well casing. The tool is drawn up the well and electronically measures the casing diameter. Caliper logs may be used to verify well depth and construction, the presence of significant hard incrustation, gaps in the casing and unstable strata in open hole wells. Soft biofouling deposits often will not be detected, as the caliper arms will cut through the deposit.

A temperature and conductivity log of the well can indicate an inflow of water through a break in the casing. It may be necessary to pump the well during logging to induce flow through damaged casing.

Fluid movement through well screens can be recorded by a specialised impeller flowmeter when the well is pumped. This can be used to determine sections of the well screen where there are either preferential or negligible flows. Casing is run inside the well to the base of the pump to prevent entanglement of the probe around the well pumping equipment. The fluid movement tool commences logging from the base of the pump to the bottom of the well. The impeller type flowmeter can become clogged with dislodged biofouling deposits.

Density logs measure the bulk density of the material external to the well casing using gamma radiation. It is useful for determining the presence of a gravel pack and although the gravel pack and casing will attenuate the signals, major aquifer lithological and stratigraphic information can be obtained.

A sonic tool generates and then measures the velocity of acoustic waves. Below the water table a cement bond log and variable density log can be obtained. In a cased hole these logs are primarily used to determine the structural integrity of the cement bond between the casing and the formation.

CCTV cameras are a valuable tool for a range of hydrogeological and drilling work. Some applications include:
- Confirming well construction details
- Determining internal casing condition due to corrosion or fouling
- Assisting in the retrieval of lost drilling equipment
- Providing geological information about fracture sizes and bedding planes to supplement geophysical logs

There are several different designs of CCTV camera viewing head. Some cameras provide separate viewing heads to look down the hole (axially) and directly at the casing wall (radially) while other cameras have a single rotating head that can pan in all directions. The problem with separate viewing heads is that the camera has to be removed from the well to change from an axial view to a radial view. This can be a problem in wells where the visibility in the water is reduced every time the camera is run into the hole. For depths up to 100 m cameras that are designed and used mainly for sewer work are adequate although they need to be thoroughly disinfected before insertion into a well. When there are over 100 m of water head those cameras with high-pressure seals designed for wells may be needed.

Black and white cameras are suitable for investigations where there is not a great deal of interpretation required from the CCTV log. This would include looking for lost equipment or determining well construction details. However a piece of casing scale can look like a casing hole with an axial viewing black and white camera. Colour cameras allow the discrimination of strata and may distinguish the fouling deposit type based on colour. Unless there is a good reason for using a black and white camera, such as camera size or cost, colour should be preferred. It is important that visual logs are just one tool used to diagnose well deterioration problems. An iron biofouling deposit may appear to impede water flow by covering a significant portion of the screen but only hydraulic testing of the well can confirm that the well yield is reduced.

The key to ensuring good quality CCTV logs is ensuring the clarity of the water in the well. The water near the surface is often murky in mild steel casing due to the enhanced corrosion around the static or pumping water levels. The removal of the pump equipment from the well can disturb soft fouling deposits on the casing creating murky water. If the goal of the survey is to view the fouling deposits intact then minimal disturbance of the well casing is recommended. This includes running the CCTV before other logs or backflushing the well by injecting clean water down the well casing either at the wellhead or at the viewing head as the CCTV camera is lowered down the well. Chemical additives (e.g. Calgon) that aggregate the particles prior to running the camera have been used with mixed success. The additive should be introduced into the well at least three days before the intended camera survey. If the survey is to identify casing damage and fouling deposits are suspected then it would be prudent to mechanically treat the well and pump it before CCTV inspection. If oil is present on the water surface this should be removed with a bailer. Detergent either put down the well or wiped on the camera lens may be effective in breaking the oil film in the vicinity of the lens.

4.7 Corrosion Data Analysis

Monitoring for corrosion can involve the use of indirect or direct measurements. The most common indirect measurements are the collection of water quality indicators (Section 4.3.1) and the calculation of corrosion predictors (Section 4.3.2). The direct methods involve the measurement of actual metal loss through the use of coupons.

4.7.1 Coupon Weight-Loss Method

The weight-loss method involves installing metal coupons either near the wellhead or in a downhole holder and retrieving the coupons after a set period of time (at least 1 month). The rate of corrosion often exponentially decreases with time (Appendix D) and a test which involves the collection of coupons at several time periods will give more accurate data on long-term rates. The results of the test give a uniform or general corrosion rate over the surface of the coupon. This data will not indicate corrosion rates attributable to pitting. A corrosion study using coupons is presented in Appendix D. This method requires attention to the coupon type, metallurgy, cleaning, weighing and exposure time as detailed in the ASTM Standards D2688 (1983), 631-72 (1972), G1-81 (1983) as well as Water Codex (1982).

4.7.2 Electrochemical Rate Measurement

Instruments are available which can measure either an electrochemical response to corrosion or the increased resistance in an electrode as corrosion occurs. These methods require trained personnel to collect and interpret the data.

5 CONTROLLING WELL DETERIORATION

Although it is occasionally possible to eliminate well fouling through well design changes it is usually a matter of controlling the impact of fouling on well performance. This is achieved through the application of adequate monitoring, treatment and operation practices and well design and equipment selection. The control of well corrosion generally involves material selection and protective coatings.

5.1 Evaluating Maintenance Options

Once the source of well deterioration has been identified a judgement must be made on the significance of the problem. This will involve determining what the current and future impacts from the problem will be on well performance indicators, what types of maintenance options are available, and what is their implementation costs.

An assessment of the current and future impact upon well performance indicators from a diagnosed problem is site specific. The unique hydraulic, chemical and microbial environment created around a well is a function of the aquifer, well design and well operation procedures. A well that produces small amounts of sand because of an incorrectly designed gravel pack is likely to be stable and not deteriorate further in the future. While pumped water which contains material from the gravel pack due to a casing rupture may get significantly worse in the future as the available gravel pack is depleted and sand from the aquifer which requires less energy to mobilise is available. It is possible that a well that produces turbidity from biofouling may show some hydraulic impairment in the future. However there is no consensus about the rate at which these processes may occur. The best management option is to rely on good monitoring records from which future impacts can be estimated from trends in the data. Where possible it is desirable to set some thresholds for action.

With well deterioration there may be control strategies based on options from corrective, design-out or operational maintenance procedures. If the main problem was an increase in the turbidity of the water from iron biofouling this could be tackled in several ways. The well could undergo chemical and mechanical treatment, use the redwater bypass as operational maintenance or a water treatment unit could be installed to treat the pumped water. If the well produces sand during well startup the well could undergo corrective maintenance by surging or swabbing to develop it, or by operational changes and inwell equipment. The choice among the options may be based on the availability of suitable technology and contractors, urgency of the required solution and cost.

5.2 Preventative Maintenance through Monitoring

To operate and maintain a well efficiently it is necessary to keep adequate records. This will allow preventative maintenance and early warning systems to indicate operational problems before major and expensive well failure occurs. It is also vital to the performance diagnosis since many problems are based on the relationship between the well design and hydrogeology. The amount of detail needed to be collected and stored in the monitoring records depends on the likelihood and importance of well failure. High capacity municipal water supply wells require more extensive record keeping than a back up domestic well. The type of data collected for records involve water quality, power usage, hydraulic, maintenance and well construction details.

5.2.1 Water Quality Records

It is desirable to collect basic water quality data that can indicate the performance of the well. The water quality parameters collected and their frequency are dependent on the budget allocated for monitoring the well, which should reflect the purpose, significance and potential problems associated with the well. Often a full water quality analysis will be undertaken after well completion. This would include both physical (pH, EC, turbidity) and inorganic water quality parameters (Ca^{2+}, Mg^{2+}, Na^+, K^+, Cl^-, SO_4^{2-}, HCO_3^-, CO_3^{2-} and alkalinity). For water supply wells a more comprehensive water quality survey should be undertaken. If there is a possibility of organic contamination from pesticides or petroleum or chlorinated hydrocarbons then a full suite of these analytes should be undertaken. A bacterial analysis of the water using bacteria that can indicate contamination (e.g. *E. coli*) is also necessary. A greater degree of judgement is required in the selection of water quality parameters that will be measured on a routine basis. These may need to meet both public health and performance diagnosis requirements. For a high yield water supply well the collection of physical and inorganic species (e.g. Fe^{2+}), turbidity and bacterial levels would be considered appropriate. While for a low yield irrigation well, water quality parameters would not be routinely collected and turbidity would be assessed visually during pump startup. The type of aquifer will also influence the water quality data collected. In an unconsolidated aquifer where the well regularly produces sand a measurement of sand production at startup would be appropriate. The frequency of these measurements can vary from daily to annually.

The onset of biofouling within a well may be evidenced by a change in water quality indicators or through the buildup of deposits on well surfaces or solid sample collection devices. However the amount of biofouling present in the well is difficult to interpret from water quality data. The sample collection devices that rely on particulate filtering (e.g. Moncell, Filters) are influenced by the frequency of the pump startup since this mobilises the biofouling deposits on the casing. Data collected from wellhead devices can be more applicable to biofilm growth in downstream pipelines while *in situ* collectors may give a better indication of biofilm growth on the well casing and screens.

5.2.2 Operational Records

This data is used to assess if there are any yield or energy impacts associated with the well performance problem. The integration of well maintenance records with the power usage and hydraulic data is particularly important when interpreting historical data. This data should be collected on a routine basis.

The details provided could include:
- Date and time of inspection
- Well water level: This will reflect the pumping or static level depending on pump operational status
- Pump operational status (on/off)
- Instantaneous discharge rate
- Cumulative discharge volume
- Pipeline pressure head
- Pump operational hours
- Instantaneous energy input
- Cumulative energy used
- Actions performed on well operation during inspection

5.2.3 Maintenance Records

Maintenance records can be the key to evaluating well performance data and determining the economic costs of well deterioration. This assists in designing cost-effective rehabilitation strategies.

The details provided should include:
- Operations on well, pump, pipelines and associated equipment
    - Date
    - Type of activity
    - Equipment repaired/replaced
    - Pump specifications
    - Pump inlet depth
    - Condition of equipment
    - Diagnostic tests performed (e.g. pump or well performance test)
    - Observations (e.g. noise, vibration)
- Economic costs of the work
    - Labour
    - Equipment

5.2.4 Well Construction Records

It is often difficult to assess the origins of well deterioration since the well casing cannot be directly inspected. Well construction records can be vital to help infer causes and identify problem areas to be further investigated. Comprehensive well construction records should contain:

- General information
  - Well identification and license number
  - Co-ordinates of the location and elevation
  - Name of the driller
  - Date of completion
- Drilling information
  - Geological log
  - Geophysical log
  - Details of any water inflows
  - Types of drilling methods and drilling mud
  - Diameter and depth of hole drilled
- Well casing
  - Type, length and thickness of casing
  - Screen type, length, diameter and slot size
  - Location and types of casing joints
  - Type and location of casing centralisers
  - Gravel pack material, size and depth
  - Cement grouting or sealing details
- Well completion
  - Development method and duration
  - Disinfection procedure
  - Water level
  - Any pumping test information
  - Any water quality information

5.3 Corrective Maintenance

There is no uniform solution that can be applied to stop well deterioration especially fouling. But time spent analysing the data available including chemical analysis and thinking about the fouling mechanisms will greatly increase the range of cost-effective and possibly innovative solutions available. Various options such as mechanical, chemical and well design changes are available to rehabilitate fouled wells. Although mechanical and chemical treatments only temporarily fix the problem and have an ongoing cost they are convenient and may be the most appropriate choice at some sites rather than well design changes which have a higher initial cost but may offer a longer lasting solution.

During rehabilitation, mechanical treatment is often combined with chemical treatment. The type of mechanical treatment and method of applying the chemical can often affect the success of the rehabilitation programme more than the particular brand of chemical used. Often the mechanical treatment selected is based on the tools, skills and type of rig available to the driller rather than the best available technology.

Only a brief description of the types of chemical and mechanical treatment is given here. Detailed information is available from other references (ADITC, 1992; Borch *et al.*, 1993; Driscoll, 1986; Roscoe-Moss, 1990).

5.3.1 Chemicals Treatment

The type of chemicals used is dependent upon the type of deposit. Inorganic chemical deposits (e.g. carbonate) only require an acid treatment while particulate deposits (e.g. fines) would need a surfactant treatment while an organic deposit (e.g. iron biofouling) would require an acid, disinfectant and surfactant phase.

There are four categories of chemicals used to remove well fouling deposits. The volume of chemical used is calculated from the required concentration in the well plus 30% of the volume of the gravel pack.

5.3.1.1 Oxidants

These chemicals are effective in breaking down organic matter. This can be bacterial cells or their associated organic based deposits ('slimes', ECP). The most popular chemical is a chlorine solution although hydrogen peroxide has been used.

A chlorine solution is commonly prepared from either sodium hypochlorite or calcium hypochlorite. When these chemicals are added to water both hydrochloric acid and hypochlorous acid are formed with the hypochlorous acid comprising the free available chlorine. The amount of available chlorine that reacts with the organic matter, hydrogen sulphide and dissolved iron and manganese in the water is the chlorine demand. The effectiveness of a given concentration of hypochlorite in water is therefore dependent on the presence of other reduced species in the water as well as any organic matter associated with bacteria. The chlorine residual is the amount of free available chlorine left when the chlorine demand of the water has been met after a specified contact period (e.g. 30 min). Since the chlorine reacts with material in the water the chlorine residual will decrease with time. It is desirable to monitor the chlorine residual during the treatment period so that its effectiveness can be determined. The recommended chlorine residuals for pathogens are 50–100 mg/L with a contact time of 0.5–2 h. For other organisms such as sulphate reducing bacteria and filamentous iron bacteria, 400 mg/L for 24 h is recommended (Driscoll, 1986). These chlorine residuals are considerably greater than the 0.1–0.5 mg/L chlorine used for water treatment. Concentrations of 1000–2000 mg/L chlorine can be necessary for biofouled wells. This makes allowance for dilution of the solution within the aquifer and the presence of deposits of organic matter in the well. Acids used in well treatment should not be used in the well at the same time as hypochlorites. After treatment the well should be pumped until there is no noticeable chlorine odour. The wastewater should not be discharged to surface water systems and may cause harm to any vegetation it contacts.

Sodium hypochlorite is often recommended for the preparation of chlorine solutions and is available in a range of concentrations. A 5.25% solution is equivalent to approximately

52,500 mg/L of chlorine. The sodium hypochlorite solution should be fresh as it does deteriorate with age. Calcium hypochlorite is also effective but a calcium hydroxide precipitate can form when the calcium ions from both the chlorine solution and the aquifer exceed 300 mg/L. Hydrogen peroxide although more expensive but decomposes in the well to leave water and oxygen. This can be useful when there are restrictions on the introduction of chemicals into a well or the disposal of wastewater is problematic.

5.3.1.2 Acids
These are used to dissolve the inorganic components (e.g. iron hydroxide) of fouling deposits and may degrade the organic matter. The most common acids used are hydrochloric (HCl), sulphamic (NH_2SO_3H) and hydroxyacetic ($C_2H_4O_3$) acids. Hydrochloric acid is the strongest and cheapest acid but hydroxyacetic acid has biocidal and chelation properties. Sulphamic acid is effective except on sulphate based deposits and has the convenience of being a dry granular material. An important consideration in the use of acids is disposing of the spent liquid waste. The excessive use of acid can cause well screen failure (Plate 8).

5.3.1.3 Surfactants
Surfactants alter the surface properties of particles and cause them to disperse which also allows treatment chemicals to penetrate further into fouling deposits. Polyphosphates are commonly used and should have some hypochlorite added (50–100 mg/L) to inhibit bacterial growth (Howsam *et al.*, 1995). Surfactants can be used to remove fines from the gravel pack but if silt and clay lenses are present these will also be dispersed and may cause turbid water.

5.3.1.4 Propriety Blends
Proprietary chemical blends should have the properties of a surfactant, acid and biocide. They offer convenience and ease of application through being ready mixed. There are no wonder chemicals that are applicable for all situations. It is likely with iron biofouling deposits that the method of application is going to ensure the success of the treatment more so than the use of a particular proprietary blend of chemicals.

5.3.1.5 Equipment Disinfection
During well construction and well maintenance programmes the microbiological conditions within the well can be changed and disinfection practices should be used. These procedures are generally based on chlorine although other biocides can be effective.

During installation all well equipment should be dusted or sprayed with a chlorine residual of greater than 200 mg/L. Deposits should be removed from equipment surfaces since disinfection is often ineffective on deposits. Water used during the drilling and drilling mud should be disinfected as they can provide a pathway for bacteria to be transmitted from the surface to the aquifer. Earthen drilling mud pits should be lined since the soil bacteria will contaminate the drilling mud. After well construction the gravel pack well casing and pump equipment should be maintained at a chlorine residual of 50 mg/L in the well for between 12 and 24 h.

5.3.2 Mechanical Treatment

In some alluvial aquifers development techniques such as overpumping and surging may not develop a well throughout the entire screen length. The upper screen becomes better developed and more permeable with subsequent development focused on these areas with the lower screen left poorly developed. Well development techniques that isolate specific sections of the aquifer for development (e.g. surge block, jetting) ensure more uniform well development. However well development techniques (e.g. surge blocks) need to be used with caution in some hydrogeological environments. Mica flakes with the aquifer apparently can orientate themselves perpendicular to the flow and decrease flow through the gravel pack. Clay lenses may break up within the aquifer causing aquifer clogging particularly if a surfactant is used.

5.3.2.1 Brushing
This is an effective method of removing soft (e.g. iron biofouling) deposits from the casing or screen. It will remove the bulk of the deposit from the well so that the amount of chemical treatment needed is reduced. The well should be pumped after brushing. Brushing may tend to smear some of the deposit into the slots and so should be used with chemicals.

5.3.2.2 Surge Blocks
Surging involves moving a close fitting block up and down within the blank casing so that water movement is induced through the well screen. It is effective in wells with short screens and uniform aquifer permeability. But in screens which are partially clogged most of the flow into the screen will be through the more permeable unclogged portions that reduces its effectiveness.

5.3.2.3 Swabbing
This involves a looser fitting block (about 12 mm clearance) than a surge block. It is operated with a reciprocating motion within the well screen. It operates by forcing water through the gravel pack around the swab. There are several configurations using either single or double swabs that can be used with water pumped into the swab or water within the well. Its action can be focused onto a particular section of screen. The double swab with water injection is generally the most efficient well development tool.

5.3.2.4 Jetting
Jetting involves injecting water air at high velocities from near the well screen into the formation. It is limited to continuous slot screens since bridge or louvre type screens deflect the jet from the formation. Its effectiveness is based on causing agitation of aquifer material within the gravel pack. When it is difficult to induce this motion the energy from the jet is rapidly dissipated.

5.3.2.5 Sonic Disruption

Sonic waves can be combined with high-pressure jets to give effective breakdown of hard carbonate type deposits. Soft deposits (e.g. iron biofouling) absorb the energy and the technique is not effective. A device of this type is the Sonar jet from the USA.

5.3.2.6 Drilling

Carbonate deposits in blank casing can be drilled out effectively. Use of just acids to dissolve massive carbonate deposits can rapidly generate dangerous amounts of CO_2.

5.3.3 *In Situ* Treatment

These treatments are designed to be used while the pump is still in the well and are not considered a substitute for the mechanical treatment methods outlined when rehabilitation of a heavily encrusted well is required. *In situ* treatment can be effective in maintaining the performance of wells subject to biofouling.

5.3.3.1 Chemical Residual

Some success is claimed by methods that maintain a continuous or intermittent chemical residual within the water. The chemicals used are generally biocides. A dry chlorine pellet dropper can be programmed to drop pellets in the well. In these systems pellet dropping should be timed with periods of pump use otherwise the chlorine concentration can buildup in the well resulting in unpalatable water or increased well corrosion. Continuous chlorination can also be obtained by injecting liquid chlorine through tubes to the bottom of the well. There is a danger that if the tube is punctured the very corrosive chlorine liquid will corrode the casing. Hydrogen peroxide can be used when there are concerns about introducing chemicals into the water since it decomposes into water and oxygen. When the water in the well is saline, chlorine generation cells have been used to produce chlorination (Forward, 1994; Gurrappa, 1996, Section 6.11).

The goal of these methods are to frequently treat the well so that either minimal or no biofouling occurs within the well. When significant buildup has occurred it is difficult to remove deposits from the well casing just by chemical treatment. Greater physical agitation can be achieved by surging the well but it is often more effective to remove the pump and use other more vigorous mechanical treatment methods (Section 5.3.2). One method to monitor the effectiveness of surging a chemical is to set up a filter or tank on the pump discharge line. The water can then be intermittently pumped into the tank during treatment. When only a small amount of biofouling deposit is being pumped out it suggests that the deposits accessible for treatment have been removed.

The application of the chemical residual technique depends to a large extent on the end use of the water. When the water is being pumped to waste (e.g. saline evaporation basins) or undergoing significant further treatment (e.g. remediation well) then there is more flexibility in the choice of the treatment chemical and the residual concentration chosen compared with water going directly to a water supply.

5.3.3.2 Chemical Backflush

If there is the capability for backflow past the foot valve in the riser pipe and into the pump inlet then treatment chemicals can be poured down the riser pipe. This procedure will adequately treat deposits in the riser pipe and near the pump. The concentration of the treatment chemical is significantly diluted when the chemical passes out of the pump inlet into the well. The well screen may be located too far away from the pump to benefit from this type of treatment. However well maintenance costs can be cut considerably when the primary biofouling problems are related to pump and riser pipe clogging. While removing the foot valve to encourage backflow will aid this type of treatment there can be problems with the pump going into upthrust during pump startup. This occurs until the riser pipe is filled and there is a head of water for the motor to pump against. A 6-mm hole drilled in the foot valve will allow some backflow. This treatment method is described in Section 6.11.

5.3.3.3 Well Seals

A well seal consists of a flexible disk (i.e. rubber) or an inflatable packer placed around the riser pipe that seals against the casing. An inflatable packer is a good way to minimise sand entry through damaged casing. Well seals can be used to separate various incompatible water qualities within the well which when mixed may cause fouling. This may be appropriate where an upper-screened aquifer is dewatered leading to oxygenated water cascading and mixing with waters from other aquifers. A well seal above the pump may minimise problems. A packer can also be used in saline wells to minimise the exposure of the riser pipe to corrosive water (Plate 9). A small hole or bypass tube should be placed in the well seal so that waters on either side can be equalised. This is important for monitoring pumping water levels. The upper section can be filled with fresh water to prevent the upflow of water below the well seal. If the main pathway for oxygen to enter the well is through the water surface within the well this method would be appropriate. However if other sources of oxygenated water are present then minimising air entry by using a well seal may not solve the problem.

5.3.4 Well Decommissioning (Abandonment)

The goal of well decommissioning is to eliminate any physical hazard from the well and to restore as far as possible the aquifer isolation that existed before the well was drilled. This is to prevent the well creating groundwater contamination from surface water entry or inter-aquifer mixing. A properly decommissioned well should be completely sealed so that the groundwater is confined to the specific zone from which it originally occurred. This requires that the vertical movement of water with the well including the annulus is prevented. There should also be a surface cap, which is mounded to prevent the ponding of water. Records should be kept of the decommissioning procedure and the new well design.

5.4 Operational Maintenance

There are modifications to the operation of the groundwater extraction/injection system that can minimise either fouling or the impact of fouling.

5.4.1 Redwater Bypass

A rapid change in water flow velocity will cause iron biofouling deposits on the pipe walls to break up and cause 'redwater'. This is particularly noticeable after pump startup and may last for up to 15 min before the turbidity reduces. During this period the water can be pumped to waste to avoid transmitting biofouling deposits and red-water through the water distribution pipelines.

5.4.2 Pump Cycling

Groundwater pumping systems that are automatically activated by demand to fill tanks may be frequently turned on and off (pump cycling). This can cause excessive sand, 'redwater' and biofouling deposit production due to rapid flow velocity changes and water oxygenation. Other effects such as increased corrosion in the casing splash zone can occur. Consideration should be given to increasing the time of the pumping cycle by decreasing the flow rate.

5.4.3 Pulsed Injection

Where chemical/microbial incompatible fluids need to be injected into the aquifer then a sequential or pulsed injection pattern has been used with success (Semprini *et al.*, 1990). This can be the case in stimulating *in situ* bioremediation where it is necessary to inject a growth-limiting compound.

5.4.4 Injected Concentration

It may be possible to change the concentration of an injected microbial stimulant to reduce the impact on well performance. When high concentrations of hydrogen peroxide are used (>7%) the chemical acts as a bactericide and it is not until the concentrations are diluted at some distance from the well that microbial activity and hence biofouling will be initiated.

5.5 Design-Out Maintenance

Minimising the impact of fouling on the performance of a well can be achieved through the careful well design and the selection of equipment, while corrosion in new wells can be minimised through adequate well design and material selection. Corrosion protection measures for existing wells are more difficult since only the pump components and above ground items are accessible for modification. The well casing is largely inaccessible.

5.5.1 Flowmeters

The well yield is usually the primary concern when operating a groundwater extraction well. It is important that the flow from the well is recorded accurately. Many flowmeters have a spindle, which rotates at a rate proportional to the flow velocity. Biofouling of the spindle changes the spindle size and will cause the flow rate to be overestimated. Significant increases in the well specific capacity with time can indicate flowmeter fouling. Other types of flowmeters, which do not use moving parts, may be more appropriate in biofouling environments (e.g. ultrasonic meters). Provision can be made to clean the flowmeter without removing it by installing access ports and isolating valves in the pipe. Treatment chemicals can be introduced into the access ports to dissolve the deposits. Inspection of the flowmeter condition should be incorporated into a regular maintenance schedule if fouling occurs.

5.5.2 Riser Pipe

Flexible riser pipe constructed from textile-reinforced elastomer can be used to replace rigid pipe in biofouling wells. The flexible riser pipe allows the pump to be rapidly retrieved from the well using a normal vehicle and a pulley rather than a drilling rig capable of withdrawing rigid pipe. This can considerably reduce maintenance costs. On pump startup or shut down there is movement in the wall of the flexible riser pipe that dislodges any deposit buildup in the pipe. This will prevent biofouling of the riser pipe but the deposits will be transported further down the flow system.

5.5.3 Well Siting

Traditionally many water well sites have been selected on the basis of water quantity without enough attention given to water quality. Well site selection in areas where fouling occurs requires an understanding of aquifer hydrochemistry. This can involve a review of construction, geological and hydrochemical data from wells in the vicinity or the collection of new data. During the interpretation of this data the method of data collection needs to be considered. Localised bands of high salinity or iron content water may be noted with a cable tool rig while they can be overlooked during the rapid drilling of an air rotary rig. Values for iron content in water can be very variable and are dependent upon the sampling methodology used. The best solution when a wellfield is to be established is to pump a test well for a given period and then evaluate whether fouling has occurred.

The location of wells close to areas where oxygenated water is likely to be rapidly recharged into the aquifer should be avoided. The oxygenated water can react with iron in the groundwater causing fouling. There may also be concerns about water quality due to contaminated runoff water entering the well. These types of recharge areas include storm water detention basins and rivers.

The access to a well site and design of a pump enclosure should be considered when frequent pump or well maintenance is required. The security fencing around the pump house should be easily removable. The pump house may have access through the roof for the withdrawal of riser pipe or insertion of well rehabilitation tools.

5.5.4 Solids Minimisation

Reducing the sand content of pumped water in an existing well may be achieved by operational changes or through the installation of inwell equipment. Operational changes involve reducing the velocity of groundwater passing through the well screen. This can involve a reduction in the flow rate during continuous pumping. During pump startup the riser pipe should also be full of water. This will minimise the high pump rates that occur until the pump–riser pipe is full and the gate valve becomes active and reduce the pump flow rate. Resleeving the well with a specially designed well screen to equalise the screen entrance velocities through the aquifer may also be effective (Pelzer & Smith, 1990). The theory behind this device is that the upper sections of the screen have higher velocities than the lower sections due to the hydraulics of well pumping. The special screens have a low slot density at the top grading to a high slot density at the bottom. Another option is to fit a sand cyclone discharge to the pump inlet. This separates the sand and water by centrifugal forces. If the sand entry is caused by a rupture of the casing the corrective action using a swagging tool, a replacement liner or the placement of an inflatable packer over the damaged section may be appropriate.

5.5.5 Screen Type

In biofouling environments well screens are often subjected to frequent mechanical and chemical treatments. High quality stainless steel (e.g. Type 316) continuous slot screen are generally the preferred choice except if biocorrosion is a problem. If severe fouling in low pH water requires the use of large quantities of acids then failure of some types of stainless steel screen can occur (Plate 8).

Plastic and fibreglass screens although resistant to the acids used for well rehabilitation may be vulnerable to mechanical abrasion from jetting and brushing. The screen open area should be maximised to reduce the frequency of yield induced well rehabilitation and increase the effectiveness of physical and chemical treatments of the gravel pack. Typical well screen open areas for a 150 mm screen with a 60 thousandths of an inch slot are steel continuous slot (40%), louvre slot (3%), bridge slot (8%), vertical mill slot (13%), plastic continuous slot (25%), slotted plastic (13%) and fibreglass continuous slot (14%) (Driscoll, 1986).

The type of screen influences the preferred development technique. The effectiveness of jetting on slotted, bridge and louvred screens can be reduced because of the lower open area causing deflection of the water jet and minimising the agitation of the gravel pack. With these types of screens a surging tool may be more effective. However the vertical rods within the continuous slot screens make it difficult to seal a surging tool against the casing and water can pass between the rods into the casing rather than into the gravel pack.

5.5.6 Well Design

If the design of the well is not correctly matched to the physical and chemical properties of the aquifer, particle accumulation and migration can occur. Often sand production problems are due to less than adequate design or completion practices. The effective well screen entrance velocity can have a significant impact upon the mobilisation of fines from the aquifer. The screen entrance velocity should in general be 0.03 m/s (AWWA, 1990) however in incrustation forming water it may be appropriate to use 0.02 m/s (IWES, 1986). To allow for screen blockage the effective screen area is usually assumed to be 50% of the design open area. A review of borehole entrance velocity design considerations is given by Howsam *et al.* (1995).

A metal screen may produce corrosion products, which cement fine particles together lowering the open area of the screen. Alternately, corrosion of the slots will allow particle entry to the well. Sand entry may also occur due to incorrect mesh selection especially in aquifers with a large variation in particle size. A well screen located in a coarse aquifer near the margin of a finer aquifer may clog or allow sand entry through the entrainment of finer material from the adjacent aquifer.

The use of water level monitoring tubes in the gravel pack can assist in the accurate definition of well performance problems if they arise. A comparison of water levels in the gravel pack and those in the well can identify whether fouling is occurring at the screen or further into the gravel pack and aquifer.

5.5.7 Pump and Pipeline Design

Surface mounted suction pumps can biofoul quickly requiring frequent disassembly and cleaning. An option may be to design the pump so that it can be hydraulically isolated with gate valves on the suction and discharge sides of the pump. Chemicals can be circulated through the pump to remove any deposits rather than removing the pump for maintenance.

At sites where biofouling is probable provision should be made for pipeline swabbing points to allow the entry of foam rubber 'pigs'. These will scour the pipeline of any deposits.

5.5.8 Water Treatment Units

Air stripping units are often used to remove volatile organic compounds from water. These units pass air through the water that can cause degassing of CO_2 and oxidation of dissolved metals. Both carbonate and iron biofouling deposits can occur. The basic air stripper types include packed towers and tray towers. Although less efficient than a packed tower the tray towers have been found less prone to iron clogging and easier to clean. Pre-treatment of the water using an oxidising filter was found to be effective in removing iron prior to air stripping (Wolf & Miller, 1989).

5.5.9 *In Situ* Aeration

The Vyredox™ system involves the injection of air into the aquifer near an extraction well, which creates an oxidised environment causing the precipitation of iron and manganese in the aquifer (Braester & Martinell, 1988; Rott, 1985) rather than in the well. This system has not been widely applied for water supply wells but may be more suited to remediation wells. Where the equipment for air sparging and soil vapour extraction are available on a site these may be utilised near a recovery well. The success in oxidising the dissolved iron will depend on how strongly the iron is chelated with organics and the ability to stimulate an aerobic bacterial population in the aquifer.

5.5.10 Cathodic Protection

Cathodic protection is widely used in water and oil industry to mitigate the effects of corrosion on pipelines, casing and storage tanks. It has been applied to well casings in oil fields (Bich & Bauman, 1995). If it has been applied to groundwater wells these cases have not been widely publicised.

Cathodic protection eliminates the current flow between various parts of a structure. It does this by providing a flow of electrons to a surface and thus creating a cathode, which does not become corroded. The amount of current needed to keep the structure in a protected or cathodic state depends on the environment such as soil resistivity and casing to soil potential.

There are two sources of cathodic protection current:
- Sacrificial anodes made of zinc, magnesium or aluminium; these can be considered as current drains
- Impressed current where an electric generator creates a direct current (DC) positive output to the anode and negative output to the well casing

Cathodic protection works on surfaces that are in a direct line with the anode. Separate systems are required for external and internal casing surfaces and for intervals of the riser pipe. Incrustation can occur on a cathodic surface, which is a consideration when designing a system for the protection of well screens.

There are difficulties associated with obtaining good electrical contact between sacrificial anodes and internal casing surfaces. Adequate surface preparation is much easier for pump columns. Cathodic protection has found most application in external casing and storage tank corrosion control.

5.5.11 Protective Coatings

Protective coatings prevent the contact of corrosive waters with a metal surface but there has been little use of protective coatings on well casings. An abrasion of the coating creates a bare spot ('holiday'), which becomes anodic to the rest of the protected casing. The corrosion rate at this spot is much greater than if there was no protective coating. Problems with the abrasion of coatings can occur during casing handling and installation. The surfaces are also vulnerable during well equipment removal and other inwell operations.

Protective coatings have been found useful to control external corrosion on riser pipes. Areas that are particularly vulnerable are around the threaded joints and near the pump–riser pipe connection. When threaded pump column is screwed up the pipe surface near the joints is often roughened up and the protective coating damaged by the tools used to grip the pipe. Adhesive tapes offer a quick and easy method to coat a pipe and are especially useful around areas that have to be periodically disassembled. An example of tape protecting the metal surface is shown in Plate 10. The threaded joints should be coated with a hardening thread compound that will exclude water from the joint reducing crevice corrosion. This is important since once a continuous pathway through the joint into the well is established then fluid jetting can rapidly enlarge the opening (Plate 5). Coaltar based products have been widely used as a coating on riser pipes although there may be problems with the leaching of hydrocarbons into the groundwater. Galvanised steel has a coating of zinc and other products that preferentially corrode and then protect the metal underneath.

5.5.12 Material Selection for Corrosive Environments

Forward & Ellis (1994) found that in saline groundwater (e.g. 20,000 mg/L) Type 316 stainless steel was generally adequate for pump equipment and screens while the more expensive Type 904L stainless was necessary when hydrogen sulphide was present. Zinc free bronze pumps can also provide good service with the option if problems arise of coating with epoxy.

In potentially corrosive environments it is important to consider well design since the well is a permanent structure and little can be done to change the design after construction. If possible inert or corrosion resistant thermoplastic, fibreglass or uPVC wells with plastic or stainless steel screens should be used. Where metal is used then galvanic corrosion can occur when different materials are used for the well casing and screens. Particularly vulnerable are short lengths of mild steel used to separate multiple stainless steel screens in deep wells. The mild steel becomes anodic to the stainless steel and corrodes. The preferred option is to use stainless steel casing between the screens. It is also desirable to use sulphate resistant cement when the upper casing intersects saline water in the strata. Since oxygen is one of the primary causes of corrosion the presence of it in the well, pump and pipelines should be minimised.

When corrosion has occurred in the casing the well can be relined with an inert material however this reduces the internal diameter of the well. An option is to place an inflatable packer onto the riser pipe next to the corrosion hole. When inflated the packer will block the entry of aquifer material and still allow easy removal of the pump upon deflation.

6 CASE STUDIES

These case studies have been selected to illustrate fouling and corrosion process in action, as well as the approaches taken to define and manage the problems.

6.1 Cost–Benefit Analysis of Well Maintenance Strategies

Source
Sutherland, D.C., Howsam, P. & Morris, J. 1996. Cost-effective monitoring and maintenance strategies for groundwater abstraction. *Journal of Water Supply: Research and Technology—Aqua* 45(2): 49–56.

This case study shows the application of cost–benefit analysis to deep wells in both Bangladesh and Pakistan. The costs associated with well management and the frequency with which they occur have been used. The outcome is an order of magnitude estimate of the costs arising from 'with' and 'without' maintenance approach to well management.

Methodology
The examples given are an application of a methodology developed for analysing the cost-effectiveness and benefits of borehole/tubewell monitoring, maintenance, and operation strategies. The chapter provides an overview of an approach, that is fully documented elsewhere (ODA, 1993a–c). The methodology incorporates both the internal technical/monetary aspects concerning the management of the groundwater abstraction system itself and external factors that incorporate consideration of water use, water users, and the environment in which the abstraction occurs.

In the assessment of cost-effectiveness the costs of management options, frequencies with which costs are incurred and well performance over time need to be considered. The examples from Bangladesh and Pakistan consider only the internal technical/ monetary factors but do not consider the impact of maintenance on well performance. This depth of analysis is suited to an initial survey where there is some empirical data on the failure rates of equipment.

The exchange rate for the examples was 1 Pound (£) which is equal to 50 Bangladesh Thaka or 50 Pakistan Rupees. The term M&M refers to monitoring and maintenance. The costs associated with well management were estimated (Table 6.1) as well as the frequencies for well management costs (Table 6.2) and the net present cost over 20 years (Table 6.3).

Table 6.1 Well management costs for production boreholes at Dhaka, Bangladesh.

Activity	Type of costs	Thaka
Monitoring	Capital costs (year 1)	50,000
Monitoring	Recurring costs (annual)	5,000
Maintenance	Recurring costs (annual)	50,000
Well failure	Non-recurrent cost	100,000
New well	Non-recurrent cost	5,000,000
Rehabilitation	Non-recurrent cost	1,000,000
Pump replacement	Non-recurrent cost	500,000
Operating costs	Average annual with M&M	1,750,000
Operating costs	Average annual without M&M	2,500,000

Table 6.2 Frequencies for well management for production boreholes at Dhaka, Bangladesh.

Activity	With M&M (years)	Without M&M (years)
Well failure	–	12
New well	25	13
Rehabilitation	14	6
Pump replacement	10	3

Table 6.3 Net present cost over 20 years for production boreholes at Dhaka, Bangladesh.

Activity	Thaka
With monitoring and maintenance	14,000,000
Without monitoring and maintenance	22,500,000

This shows that the costs are higher by a factor of 1.6 when no M&M are carried out.

Similar information was estimated for the irrigation wells in Pakistan (Tables 6.4–6.6). This shows that the costs are higher by a factor of 1.2 when no M&M are carried out.

Conclusions

The results show that M&M strategies were cost-effective in both of these environments. However the significance of the impact was affected by the different operational and maintenance requirements for the examples. The ratio of benefit from a strategy of M&M was significantly higher in Bangladesh (1.6) than in Pakistan (1.2). This reflects the groundwater environment in Bangladesh where the operational life of equipment was impacted to a greater degree by a lack of M&M. The analysis also found that the cashflow required to maintain the wells varied. In Pakistan where the operating costs were only a small proportion of the total costs the impact on cashflow of infrequent events such as new well construction was significant. The cost–benefit

Table 6.4 Well management costs for deep tubewells in Pakistan.

Activity	Type of cost	Rupees
Monitoring	Recurring costs (annual)	20,000
Maintenance	Recurring costs (annual)	25,000
New well	Non-recurrent cost	1,000,000
Rehabilitation	Non-recurrent cost	25,000
Pump replacement	Non-recurrent cost	200,000
Operating costs	Average annual with M&M	70,000
Operating costs	Average annual without M&M	77,000

Table 6.5 Frequencies for well management for deep tubewells in Pakistan.

Activity	With M&M (years)	Without M&M (years)
New well	20	15
Rehabilitation	10	10
Pump replacement	12	4

Table 6.6 Net present cost over 20 years for deep tubewells in Pakistan.

Activity	Rupees
With monitoring and maintenance	1,150,000
Without monitoring and maintenance	1,400,000

approach used in this chapter is an effective way to consider management decisions in a lifecycle analysis framework.

6.2 Improved Water Quality through Site Investigation

Source
Fujita, H., Momose, M. & Pascual, T.V. 1990. Experimental research and development study on water well construction in area where groundwater contains iron and manganese. *Water Supply* 8(3–4, Water Nagoya '89): 402–409.

The case study shows how a thorough investigation of the hydrogeology and hydrochemistry of a site can be used to improve well design and the water quality of extracted groundwater.

Background
In the rural areas of the Philippines many water wells have been abandoned due to water quality problems associated with the high iron and manganese contents of the groundwater. The study was designed to develop an evaluation methodology for these

Table 6.7 Aquifer strata.

Zone	Depth (mbgl)*	Description
I	0–2	Brown soil that is in an oxidised state
IIA	2–18	Very fine sand to silty clay. Reduced conditions with Fe and Mn concentrations of 1–2 mg/L
IIB	18–32.5	Sandy silt and fine sand with gravel. Reduced conditions with Mn at 0.5 mg/L and Fe 2–5 mg/L
III	32.5–45	Fine sand to sandy gravel. Oxidised conditions with Mn less than 0.5 mg/L and Fe 0.2–0.5 mg/L

* Refers to metres below ground level. After Fujita *et al.* (1990).

sites and to establish the relationship between water quality and substrata in Balite, Pura, in the Tarlac province about 143 km from Manila. The area is located in an inland plain of Quaternary alluvial deposits comprising clay, silt, sand and gravel. The aquifer varies from partly to fully confined.

A field survey of the water quality (pH, EC, Eh, Fe and Mn) of existing wells was undertaken. A testhole was drilled and soil samples were taken. These were analysed for grain size and colour. Colour was used to infer iron and manganese oxidation state in the substrata. The conceptual model developed for the aquifer comprised four layers (Table 6.7).

A resistivity log showed there were high porosity zones at 20, 32 and 42 mbgl. The detailed site characterisation had shown that there were reduced conditions with higher Fe and Mn concentrations at both 20 and 32 mbgl. The well screen was set between 34 and 42 mbgl with a cement grout to seal the annulus between 32 and 34 mbgl and prevent seepage from the poor quality water in the upper section of the aquifer. The water from the constructed well had a Fe concentration of 0.1 mg/L and Mn of 0.6 mg/L. These are within the limits specified for the Philippine National Standard for Drinking Water.

Conclusion

The case study shows that at some sites it is necessary to have a thorough investigation of the aquifer chemistry and to design wells based on this understanding. Other wells in the area that had not taken this into account had water quality problems and were being abandoned.

6.3 Determination of Rehabilitation Needs through a Comprehensive Evaluation Programme

Source

Puri, S., Petrie, J.L. & Flores, C.V. 1989. The diagnosis of seventy municipal water supply boreholes in Lima, Peru. *Journal of Hydrology* 106: 287–309. Puri, S. & Flores C.V. 1990. Monitoring and diagnosis for planning borehole rehabilitation – The experience from Lima, Peru. In P. Howsam (ed.), *Water wells: Monitoring, maintenance and rehabilitation*: 377–390. UK: E & F Spon.

This study demonstrates the application of a comprehensive diagnostic programme to determine the factors involved in the reduction of yield from a major wellfield. This information was then used to prioritise the wells to gain maximum benefit from any rehabilitation work. The approach may be useful as a guide elsewhere.

Background

About 45% of the water supply for Lima, Peru is obtained from over 300 boreholes. The alluvial fan aquifer is unconfined and consists of well rounded boulders, cobbles, coarse sands and lenses of fine silty sand. However the maximum yields had dropped from up to 90 L/s in the 1950s to 15 L/s in the mid-1980s. While in 1986 the water demand was $21.9 \, m^3/s$ it was projected to rise to $34 \, m^3/s$ in the year 2000. An effective plan was needed to ensure that future water demands were met.

Diagnosis Programme

It was thought that the yield had been reduced due to a combination of general aquifer yield decline, borehole deterioration and the inefficient operation of pumping equipment. The programme was planned to investigate thoroughly a representative selection of wells. Seventy wells were selected based on high fouling potential, large static water level declines, deep water tables and large saturated aquifer thickness. These wells were then subject to a series of tests (Table 6.8).

Table 6.8 Diagnostic tests conducted on the wells.

Aspect diagnosed	Tests	Data outcomes
Regional aquifer influences	Data review: Water levels, pumping rates	Historical trends
Local aquifer influences	Step drawdown test	Well and aquifer losses, well efficiency
Well condition	Verticality and alignment	Potential for well deepening
	CCTV logs	Location, length, type and condition of screens
	Geophysical logs: Gamma ray log, temperature–conductivity log	Formation properties
Well chemistry and bacteriology	pH, Eh, temperature, DO, EC, alkalinity, CO_2, H_2S, Fe^{2+}, major cation/anions, bacterial counts, corrosometer reading	Potential for corrosion, fouling processes
Pump inspection	Pipe work condition, impeller and bowl condition, motor condition, delivery valve and mains condition	Inventory and photo record
Pump house control equipment	Inspect chlorination system, switchgear and electrical system, instruments: Flowmeters, level indicators, manometers	Inventory
Pump performance	Pump testing (during step drawdown test)	Energy consumption, pump efficiency

Modified from Puri *et al.* (1989), Puri & Flores (1990).

The reliability of some of these tests was influenced by the condition of the wells. The verticality and alignment test was impacted by large amounts of fouling in the wells and other obstructions. The presence of fouling in the wells may also have masked the aquifer properties from the natural gamma logs.

This comprehensive review of the wells allowed the factors influencing the yields in the wells to be identified. Despite the poor physical condition of the wells a large number of them were operating at very low efficiencies and at higher than optimum operating costs. The poor verticality contributed to excessive pump wear. In extreme cases, undocumented casing diameters resulted in operators installing oversized pump that consequently jammed in the boreholes. The poor screen design in the older holes, very high initial pumping rates and overdimensioned pumps all contributed to fouling, aquifer overdraft, and rapid pump deterioration.

Rehabilitation Priorities
The wells were then ranked based on the likely return for any rehabilitation work needed. A high priority was given to wells with a large increase in yield for minimal rehabilitation work. The criteria used were the specific capacity from the diagnostic test, ratio of current to initial specific capacity, length of water column below the static water level, ratio of the original to present static water level and the well efficiency. A scoring system was given for each criterion with the wells requiring least work having the highest score.

The information from the diagnostic programme was also used to evaluate the cost–benefits associated with any proposed rehabilitation work. The well loss component of the step drawdown test analysis was used to estimate potential yield improvements from rehabilitation. The increased specific capacity associated with deepening wells was also considered. This information was integrated into a mathematical model of the regional aquifer. The combined results from the diagnosis recommendations and the modelling allowed a reasonable basis for anticipated benefits from rehabilitation.

Conclusions
The structured and thorough approach to understanding the causes of declining well yields has allowed a rehabilitation strategy to be proposed which has a high probability of producing the benefits anticipated from the rehabilitation work undertaken. It was estimated that a 56% increase in yield could be obtained from the diagnosed wells if they were rehabilitated. It was also estimated that by restoring the pumps and motors to an operating efficiency of 60% the cost of the work would be recovered in three years through lower operating costs.

6.4 Borehole Deterioration in a Micaceous Aquifer

Source
Bakiewicz, W., Milne, D.M. & Pattle, A.D. 1985. Development of public tubewell designs in Pakistan. *Quarterly Journal of Engineering Geology* 18: 63–77. Blackwell, I.M.,

Howsam, P. & Walker, M.J. 1995. Permeability impairment around boreholes in micaceous aquifers. *Quarterly Journal of Engineering Geology* 28: S163–S175.

A high number of tubewells installed in micaceous aquifers in Pakistan have had both short- and long-term performance deterioration. Although the causes for the decline have not been conclusively shown, a combination of laboratory and field research along with a review of monitoring data have been effectively used to identify key processes.

Background

Within Pakistan, many thousands of tubewells have been drilled into an alluvial aquifer which contains up to 70% mica. In this environment there have been a greater than expected decline in well performance. Many of the wells have completely failed after 5–10 years of operation. A study of 38 wells found an average deterioration in their specific capacity of 30% with some wells declining by 80%.

A number of possible causes of well deterioration were considered:
- Chemical incrustation of the screen slots, gravel pack or formation/gravel pack interface
- Migration of fines
- Chemical alteration of mica
- Physical re-orientation of fines/mica

Each of these potential causes were examined using relevant laboratory and field research as well as a review of monitoring data.

Chemical Incrustation of the Screen Slots, Gravel Pack or Formation/Gravel Pack Interface

Attempts were made to correlate hydrochemical properties of the well water with well performance deterioration. Relationships between pH, Eh and dissolved gases along with specific capacity data were not found. However the analysis was hampered by the lack of wellhead measurements since these properties can vary significantly from the laboratory data used in the analysis.

At selected sites the underwater CCTV logs of well screens and visual examinations of extracted wells did not show any evidence of chemical incrustation. These pieces of evidence suggest that chemical incrustation was not a problem in these wells.

Migration of Fines

To evaluate whether mechanical clogging of the gravel pack was likely, a relationship between the degree of well performance deterioration and various grain size characteristics was sought. Statistical examinations of the specific capacity of selected wells and the median and effective particle sizes of the aquifer showed no relationship. Some laboratory studies using samples from Pakistan showed that fines migration occurred in the first 20 min of pumping and comprised just 0.03% of the sample. This suggested that fines migration was not a key process in the long-term deterioration of well performance in these wells.

Chemical Alteration of the Mica

If the main cause of well performance were chemical then a relationship would have been expected between pumped water quality and well performance deterioration. The pumped water salinity was divided into various classes based on initial EC and its subsequent rise or fall with time. The results from Khairpur involving the analysis of 380 wells using 10 years of data was inconclusive. The analysis from the 180 wells at South Rohri over a three-year period showed a marked decline in specific capacity of a salinity group. It was considered possible that potassium in the mica could be exchanged with cations in the groundwater casing the mica to swell and fray. Laboratory tests showed that significant amounts of potassium can be removed from the mica by leaching with saline water. Experimental modelling showed that mica expansion could have an impact on hydraulic conductivity in the aquifer of 20–99%.

Physical Re-orientation of Fines/Mica

A unique feature of the aquifer was the horizontal stratification caused by mica flakes which settled in a horizontal direction in the aquifer. This caused a high degree of hydraulic anistrophy due to the lower vertical permeability of these fine-grained lenses.

A detailed study was undertaken at an abandoned well. This involved examining the well screen both *in situ* (with a CCTV camera) and after extraction. The borehole was also examined after casing had been extracted. It was observed that there was mixing between the formation and the gravel pack. In other places the gravel pack was absent. Voids in the gravel pack may have provided an opportunity for the formation to slump and mix with the gravel pack. It was thought that a major cause of the well deterioration was a physical re-orientation of the mica flakes from a horizontal position due to formation damage causing a reduction in horizontal permeability near the well. Laboratory and modelling results show that the scale of well performance decline from near borehole permeability damage is consistent with that observed in the field.

Summary

The tubewells in Pakistan have a history of rapid decline in well performance after commissioning. An investigation process over a number of years using statistical analysis of monitoring data, field investigations, validated by laboratory and modelling studies have identified several key processes that are likely to play a role. The borehole drilling and construction practices may have allowed formation and formation/gravel pack mixing leading to a reduced permeability near the borehole. A change in the pumped water quality may have altered the mica in the formation causing reduced permeability.

6.5 Microbially Enhanced Corrosion of Well Screens

Source
Elshawesh, F., Elmendelsi, T., Elhoud, A., Abaan, E. & Elagdel, E. 1997. Micro-biologically influenced corrosion causes failure of Type 304 water well screens. *Materials Performance* 36(6): 66–68.

Stainless steel well screens were found to fail after two years in a freshwater aquifer. The case study demonstrates the need for best practice in well development and monitoring.

Background
A number of water wells were installed in a freshwater aquifer at a depth of 500 m. The well screens were Type 304 stainless steel. The wells were left stagnant for two years but upon pumping a large quantity of different sized sands and gravel came out of the well.

Investigations
An underwater CCTV camera log showed the screens had failed at several locations. Some well screens were removed from the borehole and examined both visually and microscopically. Two types of corrosion attack were evident.

One type involved isolated pits along the rods while the other appeared to be crack-like ditches. In some places over 50% of the cross section of the rods had been corroded. This had led to structural failure of the screen. The wires of the screen were not significantly corroded.

Samples of both the groundwater and deposits from the affected areas were taken. The deposits on the metal surface were analysed and showed the presence of chloride, iron, sulphate reducing bacteria and iron bacteria. The water analysis found chloride ranged from 105 to 470 mg/L and sulphate was around 235 mg/L. Both iron bacteria and sulphate reducing bacteria were also detected in the water.

Conclusion
The nature of the corrosion pits and the deposit analysis suggested both microbial influenced corrosion and chloride assisted pitting attack could have occurred.

It was thought that bacterial contamination from drilling fluids and the lack of chlorination during well drilling, development and operation may have contributed to bacterial activity around the wells. The two-year period where the wells were not used created stagnant water which was prone to the establishment and growth of corrosive microbiological colonies.

It was suggested that regular monitoring of wellfields for water quality is essential to identify deterioration of water quality, bacterial contamination and the effectiveness of chlorination.

6.6 Fouling Caused by Well Casing Corrosion

Source
McLaughlan, R.G., Knight, M.J. & Stuetz, R.M. 1993. *Fouling and corrosion of groundwater wells: A research study*. Research Publication 1/93, National Centre for Groundwater Management, University of Technology, Sydney, 213pp.

The study illustrates the well deterioration that occurred as a result of casing corrosion. Both water analysis and chemical equilibrium modelling can be useful in determining the causes of processes involved in well deterioration due to carbonate fouling.

Background
In 1967 an artesian well was drilled at Karumba, North Queensland. It penetrated sand and shales and was constructed with a slotted interval from 720 to 747 m depth resulting in groundwater with a temperature of 57°C. The pump inlet was set at around 55 m.

Numerous mechanical problems occurred with the line shaft pump during well operation. It was suspected these were related to well alignment and the high water temperature. After 1971, the deposition of calcium carbonate reduced the pump service interval to less than one month. The calcium carbonate buildup also reduced the internal diameter of the riser pipe, leaving only a small opening for the pumped water. The deposit extended from the surface through 48 m depth. The incrustation shown in Plate 11 occurred over a four-month period. In 1976 the well was relined to a depth of 100 m and although there was some carbonate deposition the pump service interval was increased to at least nine months.

The presence of a thin upper saline aquifer was noted after a nearby well had been drilled with a cable tool rig. It was not noticed in the drilling of the deep well, possibly because other drilling techniques were used.

Methodology
The well fouling deposits were analysed to determine their composition using XRF and their mineralogy using XRD. The samples were predominantly calcite (calcium carbonate).

Water samples from the upper and lower aquifers were analysed for ionic composition. Samples from both the upper and lower aquifers as well as composites of the two waters were evaluated by chemical equilibrium modelling for their tendency to precipitate. This attempted to understand what the key processes were that may cause these samples to precipitate. The results showed that only the composite water samples and those from the lower aquifer had a tendency to precipitate calcium carbonate.

A shallow saline aquifer is suspected of corroding the upper casing and mixing with the carbonate rich groundwater from depth causing precipitation. The chemical equilibrium modelling found the amount of precipitate was greater for the composite sample than the sample from the lower aquifer. This may explain why even after relining to reduce any inflow from the upper aquifer there was still some deposit buildup.

Conclusion

This case study shows that the massive carbonate deposition was due to the mixing of incompatible waters while the minor fouling that occurred after relining was due to pressure (CO_2 degassing) and temperature induced change in the groundwater during pumping. It also shows the impact that highly localised aquifers can have on well performance.

6.7 Aluminium Biofouling

Source

McLaughlan, R.G., Knight, M.J. & Stuetz, R.M. 1993. *Fouling and corrosion of groundwater wells: A research study*. Research Publication 1/93, National Centre for Groundwater Management, University of Technology, Sydney, 213pp.

This case study shows that detailed analysis of fouling deposits and *in situ* water chemistry can be useful in determining whether fouling was caused by the mixing of different water chemistries within a well. It also shows the variability of water chemistry that can occur within an aquifer.

Background

At the Mirrabooka wellfield in Western Australia, aluminium biofouling deposits (Plate 12) have caused a significant maintenance problem in at least eight wells. The wells are 30–50 m deep and draw water from sediments comprising sand with minor limestone and silt lenses.

Methodology

A chemical analysis of the deposits from a well was undertaken to determine the composition of the deposits. This information can be extremely useful in inferring the key processes which cause a particular deposit to form. The chemical analysis of the deposits shows that aluminium is the dominant element within the deposit (Table 6.9). The LOI_{400} shows that there is a very high microbial component to the deposit, which may include the sulphur related bacteria. The small difference between the LOI_{400} and LOI_{1050} and small Ca value suggests that there are few carbonates in the deposit.

Water samples were also taken from the well at the wellhead and at various depths with a packer (Table 6.10).

Table 6.9 Chemical analysis of well deposit from well M340.

Si	Al	Fe	S	Mn	Ca	LOI_{400} (%)	LOI_{1050} (%)
0.78	14.2	0.05	0.42	0.0	0.04	60.7	68.71

Elements are % XRF oxide recalculated for elemental form.

Table 6.10 Hydrochemical analysis of well M340.

	Sample location			
	Wellhead	23.5–24.5 m	31.5–32.5 m	40.5–41.5 m
Degree of biofouling on screen		Heavy	Light	Nil
pH	5.43	7.31	7.28	7.31
EC (mS/m)	50	52	52	53
Fe	3.4	0.95	1.0	1.2
Al	2.0	0.11	0.078	0.115
Na	55	32	34	36
Ca	9	60	65	65
Mg	9	4.8	5.5	5.5
Cl	82	58	60	60
SO_4^{2-}	50	4	9	9
Alkalinity (meq)	0.38	3.3	3.4	3.3
Sulphide	1.0	0.5	0.5	0.5

All units except pH are mg/L unless otherwise stated.

The data shows that the water from individual depths is similar to each other but vastly different from the water collected at the wellhead during pumping. The higher pH waters from the various layers must be mixed with the lower pH aluminium rich water from another part of the aquifer that was not sampled. Since aluminium solubility is reduced at higher pH (Appendix A2), this mixing of different waters has caused the precipitation of aluminium. The distinctly different ionic composition of the wellhead and sampled waters (e.g. Na, Cl) show that there are two different waters within the aquifer and that the lower pH water contributes most of the flow to the well.

Conclusion

Although a cause for the precipitation process was established, there was no solution available to stop the problem. Changes to the well design at each of the sites would have required detailed packer testing down the entire screen length to establish the location of the hydrochemistries for the aquifer. These sections would then need to be blocked off. Since this type of survey would have been required for at least eight fouling wells in the wellfield it was considered impractical. A regular (2–8 monthly) maintenance schedule was implemented. The well fouling deposits were found to be more amenable to dissolution by sodium hydroxide than acid treatment.

6.8 Impact of Variations in Aquifer Hydrochemistry on Iron Biofouling

Source
McLaughlan, R.G., Knight, M.J. & Stuetz, R.M. 1993. *Fouling and corrosion of groundwater wells: A research study*. Research Publication 1/93, National Centre for Groundwater Management, University of Technology, Sydney, 213pp.

This case study shows how an investigative approach was used to determine the key cause of biofouling in a wellfield. The occurrence of localised hotspots of dissolved iron was found to cause well biofouling.

Background
To help reduce waterlogging and land salinisation from irrigation a number of large-scale dewatering schemes have been set up in Australia. At Wakool 47 pump sites discharge shallow saline water to a large evaporation basin. Iron biofouling in some of those wells (Plate 2) have affected pump performance and increased maintenance costs. The pump sites have one or two wells connected to a single suction pump above ground. At sites with two wells one well may be biofouled and the other well 30 m away is unaffected. It was a concern that the affected well may contaminate the clean well increasing the biofouling problem across the site.

Methodology
It was decided to examine whether there were differences in the microbiological quality of the water across the whole site. Water samples from around the site showed that the aerobic heterotrophic bacterial numbers were not significantly different between the fouling and non-fouling wells and therefore this was not the controlling factor in whether the wells fouled.

It was found that the wells that were affected by fouling had a higher dissolved iron concentration than those that were unaffected. A survey of the iron concentration in water samples from over 61 wells and piezometers at the site could not relate the iron to a specific salinity or depth in the aquifer. This suggested that the occurrence of iron in the aquifer was very localised rather than following a regional trend. Six piezometers were drilled in a circle around one biofouling well at a distance of 1 m. Water sampling showed that one piezometer had iron concentrations of greater than 25 mg/L while the other piezometers showed no iron. CCTV camera logs in the well showed fouling was only on the side of the casing where the high iron concentrations were located.

Hydraulic evaluation of the fouled wells showed there was no decline in the well performance due to fouling however there were declines in the pump performance and significant costs associated with maintaining the performance of the pumps.

Conclusions
At these sites, the ideal location of a well needs to consider the aquifer hydrochemistry as well as physical infrastructure support (piping, power, access) and location to maximise regional drawdown. Given the shallow nature of the existing wells there were few options with regard to well design that could be made to decrease the fouling at the sites where this occurred. Since the water is already saline and pumped to waste, the most appropriate strategy would be a chemical backflush or chemical residual system.

6.9 Aluminium Fouling Caused by the Mixing of Different Groundwater Chemistries

Source
McLaughlan, R.G., Knight, M.J. & Stuetz, R.M. 1993. *Fouling and corrosion of groundwater wells: A research study.* Research Publication 1/93, National Centre for Groundwater Management, University of Technology, Sydney, 213pp.

This case study shows how chemical analysis of deposits and selective sampling from various depths in the aquifer, were able to identify the key processes causing fouling.

Background
The Mineral Reserve Basins (Victoria, Australia) are dry lakes, which were investigated for the storage of saline water. Around the basin 13 wells up to 50 m deep were constructed as interception wells to contain seepage. During pumping tests of the wells a white precipitate formed in the discharge lines (Plate 13) and at the pump outlet of five wells with fouling of the well screen on only one of those wells and an observation well.

Methodology
A chemical analysis of the deposits were undertaken to give an insight into the types of waters and processes that may have been involved in the formation of the deposits. Samples were taken from various parts of the well system in case different processes were acting in different parts of the system. The results showed that aluminium and to a lesser extent sulphur were the dominant elements precipitated from the groundwater (Table 6.11). The presence of Na, Cl, Ca and Mg is largely due to the high salinity water trapped within the deposit during sampling. The higher silica concentrations may be due to aquifer particles trapped in the deposit.

An investigation was then taken to selectively sample groundwater from different depths in the aquifer to identify any problematic water chemistries. This showed that there were significantly different water chemistries present at various depths. The upper aquifers around 25 m depth had a pH of 3.6 and aluminium present at 440 mg/L. The low pH water had presumably leached aluminium from the clays and kept it in a dissolved form. The water from 45 m depth had a pH of 6.5 and negligible aluminium.

Table 6.11 Chemical analysis of samples from well P1.

Sample	Al	Fe	Ca	Mg	Na	SO$_4$	Si	Cl
Well screen	35	0.7	0.1	0.3	2.0	0.7	8.6	–
Discharge pipe	22	–	0.1	0.1	1.3	8	2	5
Pump outlet	14	–	0.4	1.3	9	12	5	17
Pump outlet	28	0.2	0.1	0.4	6.0	14.8	1.7	7

Elements are % XRF oxide recalculated for elemental form.

Conclusions

In wells that were screened through both aquifers there were considerable deposits within the wells. This is likely to be caused by the deeper high pH water mixing with the shallow aluminium rich water causing precipitation. In the wells that were only screened in the upper aquifers the precipitation generally occurred in the discharge lines where the water was discharged to the surface. The dominant precipitation mechanism with water from the upper aquifer was thought to be aeration and degassing of the water causing an increased pH and subsequent precipitation.

Due to these operational problems and political reasons the scheme was abandoned. It probably would have been possible to design wells which selectively extracted the water from particular depths and then attempted to minimise the changes in water chemistry during pumping.

6.10 Fouling of Groundwater Treatment Recovery Wells

Source

McLaughlan, R.G., Harwood, R., Hasselgrove, K. & Stuetz, R.M. 1994. Fouling of groundwater injection/recovery systems, *IAH/IEA Water Down Under '94 Conference Proceedings, Adelaide, November 1994 Preprints,* Vol. 2, Part A, 203–208.

This study also shows how the use of chemical analysis of the fouling deposits and groundwater at a site lead to a better understanding of the processes causing fouling. Based on this understanding, strategies to minimise mixing of different groundwater at the site have led to a decrease in well fouling.

Background

At a mineral reprocessing plant alkaline residues were stored in evaporation basins. Historically where a defect occurred in the basal clay liner the alkaline liquor could seep into the underlying groundwater system. A network of pumping wells were installed around the residue area to remove the contaminated groundwater. These recovery wells have a significant fouling problem which requires a frequent maintenance programme.

Methodology

The groundwater across the site was characterised. The liquor in the evaporation ponds had a pH of 12, was reducing, had low levels of dissolved iron, calcium (4 mg/L) and magnesium (0.02 mg/L) but high levels of sulphide (9 mg/L) and carbonate. The sediments underlying the evaporation basin which are rich in organic material react with the hydroxides from the liquor to form a dark brown coloured groundwater plume that can have high levels of iron dissolved and chelated with organics as well as in colloidal form. The natural groundwater mostly has a pH around 7 and flows through calcareous sediments resulting in moderate levels of calcium and magnesium (70 mg/L total). Chemical equilibrium modelling was used to establish the tendency for each of these waters as well as a mixture of them to precipitate. It was found that when these waters mix there is considerable potential for precipitation.

The fouling deposits from a number of different wells were chemically analysed. This was to try to establish what the key causes for fouling were and whether these were the same across the site. There are three predominant types of deposits present: iron hydroxides and sulphides with minor Ca and Mg carbonates, and predominantly Ca and Mg carbonates with some aluminium hydroxides. The composition of the fouling deposit from a particular well is dependent upon the degree of liquor/groundwater mixing in the well and the composition of the plume liquor and groundwater at that point. There are variations in the natural groundwater composition beneath the evaporation basins which complicates the exact calculation of background water chemistries.

Conclusion
The degree of mixing between the liquor and the groundwater affects the mass of precipitate formed. This affects the number of weeks each recovery well is in operation before the clogging of the screen and subsequent drop in flow rate necessitates rehabilitation. To minimise the amount of carbonate fouling it is necessary to increase the percentage of liquor pumped from a recovery well as the liquor by itself has little potential for carbonate fouling. However the relationship between the liquor and iron levels in the groundwater has not been well enough established to pick the optimal liquor concentration to minimise iron fouling. Some reduction in well scaling has been achieved at the site by grouping the extraction wells into zones of high and low contamination which effectively reduces the liquor mixing with groundwater. In the worst scaling bores, the submersible pumps have been replaced by air operated pump since these do not suffer overheating problems from fouling of the submersible motor. The excessive use of acid to dissolve the fouling deposits in the very high pH water in the wells has caused well screen failure (Plate 8).

6.11 Well Treatment Using an *In Situ* Well Chlorinator to Reduce Iron Biofouling

Source
Forward, P. 1994. Control of iron biofouling in submersible pumps in the Woolpunda salt interception scheme in South Australia. *IAH/LEA Water Down Under '94 Conference Proceedings, Adelaide, November 1994 Preprints,* Vol. 2, Part A, 169–174.

This study shows how the use of pump and aquifer tests were able to identify the cause of well deterioration to be fouling of the pump–riser pipes rather than the well screens. This allowed an innovative treatment method to be used that stopped the biofouling problem.

Background
At the Woolpunda Salinity interception scheme (South Australia) saline water is intercepted by 49 wells and pumped to a saline disposal pond. It was found that biofouling

affected 42 wells. In the worst impacted wells the flow rate would drop by 45% after 50 days of operation.

Methodology
A detailed performance evaluation of the pump and aquifer was undertaken to pinpoint the exact location of biofouling deposits which lead to well performance decline. These tests showed that although biofouling of the well screens did occur the decline in performance was predominantly due to clogging within the pump–riser pipe.

The technology used for chlorinating swimming pools was adapted for use in wells to treat biofouling. The chlorination system uses a free chloride generator which works by using energised chlorinator electrodes to convert the chloride in the saline ground-water to free chlorine which acts as a biocide. The generation of free chlorine has been found effective in swimming pools where there is an EC of around 6000 mS/m and at this site with a chloride concentration of 12,000 mg/L. The chlorinator electrodes are housed in the wellhead and once a day the pump is stopped and the electrodes acti-vated. A small back flow of water is allowed past the electrodes and down the well, which creates free chlorine at 4 mg/L.

Conclusion
The chlorination system has been operating for over two years without any decline in performance attributable to biofouling.

6.12 Aquifer Overexploitation Linked to Increased Well Corrosion

Source
Ceron, J.C. & Pulido-Bosch, A. 1996. Groundwater problems resulting from CO_2 pol-lution and overexploitation in Alto Guadalentin aquifer (Murcia, Spain). *Environmental Geology* 28(4): 223–228.

This study shows how a link was made between increased drawdown levels in an aquifer and a change in the water pumped from wells in the area. The increased levels of salinity and CO_2 have led to increased well corrosion.

Background
The aquifer of Alto Guadalentin is located in the southeast of the province of Murcia, Spain. The aquifer is detrital and composed of conglomerates, sand, silt and clay. It overlies a complex substratum including metamorphic units.

Since 1960 the demand for water in the area increased as pumping equipment improved and crops of enormous economic advantage could be developed. The increased pumpage resulted in piezometric levels in the aquifer dropping from initial levels of 30–60 m to between 150 and 250 m depth. There was also a significant increase in the salinity and carbon dioxide content of the water. The pumped water has caused rapid corrosion of well equipment and pipelines, cavitation during pumping and precipitation of carbonates due to degassing.

Methodology

The ionic composition of water from wells showed there was a direct relationship between waters with higher concentrations and those coming from the rocks of the deeper substratum. At a well which had been abandoned an increase was found between increased piezometric level depth and the conductivity and bicarbonate ion content of the water through time. Bicarbonate ion was used as an indicator of dissolved CO_2.

Conclusion

It is thought that the increased rate of the deterioration of well equipment from corrosion is related to the overexploitation of the aquifer. Falling water levels are causing well owners to tap the deeper more corrosive waters to maintain their well yields. There may also be migration of the deeper more saline and CO_2 rich water into the shallow aquifer from the deeper metamorphic strata due to the extensive faults in the substratum underlying the shallow aquifer.

APPENDICES

APPENDIX A: WATER CHEMISTRY BACKGROUND

A1 Iron Species

In geological environments many rocks have minerals containing iron. When these minerals are weathered the iron is released which can then be precipitated in sedimentary rocks in a variety of solid forms. Under reducing conditions with sulphur present, iron sulphide species (e.g. FeS_2) may occur while if there is less sulphur then iron carbonate (e.g. $FeCO_3$) can form. In oxidising environments a range of ferric oxides (e.g. Fe_2O_3), oxyhydroxides (e.g. $FeOOH$) or poorly crystalline ferric hydroxides (e.g. $Fe(OH)_3$) can form (Fig. A.1).

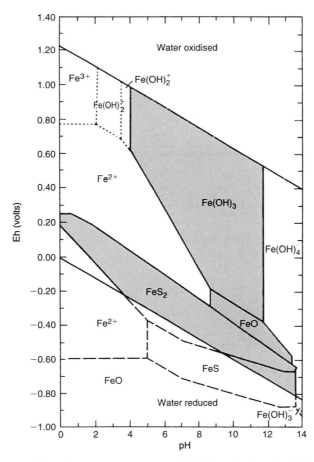

Figure A.1 Solid and dissolved forms of iron as a function of Eh and pH at 25°C and 1 atm. Activity of sulphur species 96 mg/L as SO_4^{2-}, carbon dioxide species 61 mg/L as HCO_3^-, and dissolved iron 56 μg/L. Shaded area represents iron precipitates. (After Hem, 1989.)

The amount of ferrous iron (Fe^{2+}) dissolved in the groundwater is related to the amount of solid iron deposits present in the aquifer and depends mainly on the redox and pH conditions around these sediments. The most common form of iron in groundwater is the ferrous ion (Fe^{2+}). Within the usual pH range of groundwater (pH 5–9) at relatively low Eh (below 0.2 V and above -0.1 V) considerable ferrous iron may be maintained in solution. Relatively small changes in Eh or pH can cause large changes in iron solubility (Fig. A.2).

Groundwater may contain soluble humic and fulvic acids, which have reducing and complexing properties. Complexes can occur either with ferrous iron or ferric iron. Solubility diagrams do not take into account chelation as the effects are largely not quantified. Chelation often occurs at sites with a high TOC loading such as many sites contaminated with organic chemicals. The effect of the chelates is that iron is more

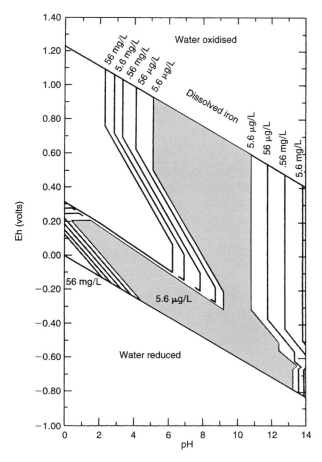

Figure A.2 Equilibrium activity of dissolved iron as a function of Eh and pH at 25°C and 1 atm. Activity of sulphur species 96 mg/L as SO_4^{2-}, carbon dioxide species 61 mg/L as HCO_3^-, shaded area represents iron precipitates. (After Hem, 1989.)

stable as a dissolved species than it would otherwise be if the chelates were not present. The presence of dissolved organic carbon (humic and fulvic acids) at greater than 3 mg/L required up to eight times the amount of oxidant to cause precipitation as that for uncomplexed iron (Knocke *et al.*, 1990).

The rate of iron oxidation by oxygen is considered relatively rapid compared with the oxidation rate for other elements. Above a pH of 5 the rate of oxidation increases by a factor of 100 for each increase of a pH unit. At a pH of 6.8 in an aerated solution the half time for the reaction is about 18 min (Hem, 1989). However the residence time for groundwater in the near well environment where the mixing of oxygenated and reduced water could be postulated to occur is considerably shorter than this. Bacteria can increase the rate of this oxidation and so help create the conditions necessary for iron fouling in groundwater extraction systems.

A2 Aluminium Species

Many silicate igneous rock minerals contain aluminium that upon weathering form sedimentary aluminium enriched minerals (e.g. clays). Although there is an ample supply of aluminium in most sediments it is one of the least mobile species because of its low solubility (Fig. A.3). Under natural groundwater pH ranges the aluminium solubility is less

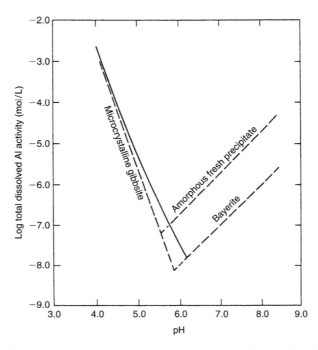

Figure A.3 Equilibrium activities of free aluminium for aluminium hydroxide (dashed lines) and calculated activity of $Al^{3+} + AlOH^{2+}$ (solid line). (After Hem, 1989.)

than a mg/L but in more acidic waters (e.g. acid mine drainage) or within isolated low pH zones in an aquifer the aluminium concentrations can be much greater (hundreds to several thousand mg/L). Water analysis that report aluminium concentrations may include particulate matter since aluminium hydroxide particles (0.1 μm diameter) can have considerable physical and chemical stability (Hem & Robertson, 1967).

Aluminium can occur in a number of dissolved forms in groundwater. Below a pH of 5, Al^{3+} is the dominant aluminium species while at higher pH values a variety of hydrolysis species dominate. Complexes may also occur with fluoride, sulphate, nitrate and with organic ligands. The extent of the complexation depends on the groundwater environment. Aluminium can form a number of different aluminium hydroxide solids as well as aluminium sulphate and hydroxysulphate minerals (Nordstrom, 1982).

A3 Inorganic Carbon Species

The carbonate–carbon dioxide system plays an important role in groundwater equilibria.

It comprises:
- Gaseous CO_2 as $CO_2(g)$
- Carbonic acid, $H_2CO_3^*$ containing
    - $CO_2(aq)$
    - $H_2CO_3(aq)$
- Bicarbonate ion as HCO_3^-
- Carbonate ion as CO_3^{2-}
- Carbonate containing solids (e.g. $CaCO_3(s)$)

These species are linked by the following equations:
$$CO_2(g) \Leftrightarrow H_2CO_3$$
$$H_2CO_3 \Leftrightarrow H^+ + HCO_3^-$$
$$HCO_3^- \Leftrightarrow H^+ + CO_3^{2-}$$
$$CO_3^{2-} + Ca^{2+} \Leftrightarrow CaCO_3(s)$$

Through the precipitation of carbonate compounds and the release of CO_2 the carbon dioxide–carbonate system can buffer natural water systems against gross changes in pH. These dissolved carbonate–carbon dioxide species are the dominant contributor to the alkalinity of natural groundwater. The alkalinity is the capacity of a water to react with and neutralise acid. In contaminated groundwater certain organic species may contribute to the alkalinity of the water and care should be taken when interpreting the carbonate–carbon dioxide species from this data.

The equilibrium distribution of the aqueous species in dilute solutions is dependent upon the solution pH (Fig. A.4). At a higher pH the carbonate–carbon dioxide species are more likely to occur as carbonate ions while at a low pH carbonic acid and gaseous carbon dioxide are more likely. When evaluating the potential of carbonate fouling

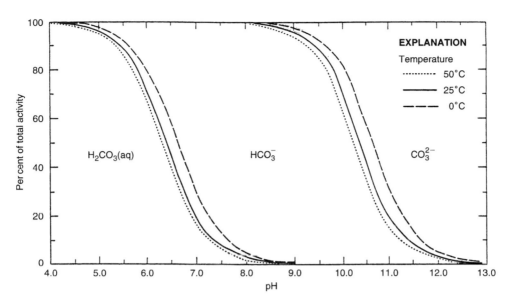

Figure A.4 Percentages of dissolved carbon dioxide species activities at 1 atm pressure and various temperatures as a function of pH. (After Hem, 1989.)

from a water of a given alkalinity it is the higher pH waters that will contain the most carbonate ions available for precipitation. The tendency of water to precipitate calcium carbonate can be calculated from the LSI. Calcium carbonate is often reported as a well fouling deposit while the magnesium bearing carbonates (e.g. dolomite, magnesite) are not often found. Although the chemical conditions may be suitable (thermodynamically) for precipitation of the magnesium bearing carbonates other factors (e.g. kinetics) can inhibit their crystallisation at ordinary temperatures (Deer *et al.*, 1966). Care needs to be taken in applying geochemical equilibrium modelling packages to ascertain whether fouling will occur based on water chemistry since these packages often do not take kinetics into account.

A4 Sulphur Species

The redox conditions within an aquifer exert a strong control over sulphur since it can exist in a range of ionic states from S^{2-} to S^{6-}. In groundwater sulphur commonly occurs in the dissolved form as sulphate but under reducing conditions the sulphide species (e.g. H_2S, HS^-) become more stable (Fig. A.5).

There are also short-lived sulphur compounds (e.g. thiosulphate ions) that are not represented in the equilibrium diagram which can contribute to the total sulphur species present. Bacteria are involved with sulphur species transformations primarily through sulphate reduction and sulphur oxidation. The sulphur species can be involved in a number of minerals that occur in well fouling deposits. When ferrous iron and sulphide

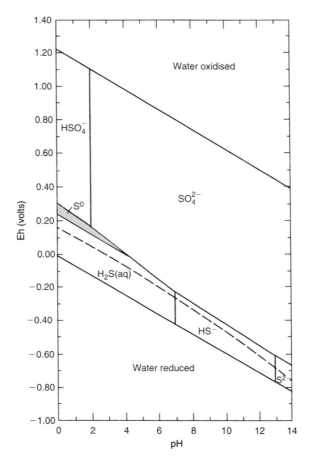

Figure A.5 Distribution of sulphur species at 25°C and 1 atm. Total dissolved sulphur activity 96 mg/L as SO_4^{2-}. (After Hem, 1989.)

are present then iron sulphide will form due to its low solubility. Sulphate can occur in aluminium hydroxysulphate minerals that have been found in well fouling deposits (Section 6.4). Sulphate can also form deposits with calcium, barium and strontium with these types of deposits reported from oilfield injection wells during seawater flooding (Jones, 1988) and in remediation wells (Hodder & Peck, 1992).

A5 Manganese Species

Manganese can exist in a range of ionic forms from Mn^{2+} to Mn^{4+} and therefore like iron and sulphur is dominated by redox reactions. In groundwater manganese is most likely to occur in the dissolved form as Mn^{2+} since Mn^{3+} is generally unstable unless under strongly acid conditions (Hem, 1989). Dissolved manganese does not favour forming complexes with dissolved organics unlike iron (Knocke *et al.*, 1990).

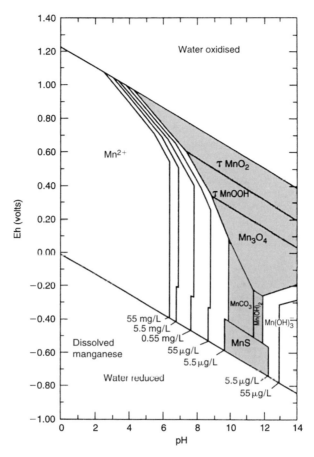

Figure A.6 Distribution of manganese as a function of Eh and pH at 25°C and 1 atm. Activity of sulphur species 96 mg/L as SO_4^{2-} and carbon dioxide species 61 mg/L as HCO_3^-.

Manganese is generally more stable to oxidation than iron that is reflected in the zonal precipitation patterns observed where oxygenated water has been injected into a reduced aquifer. Manganese was found to precipitate in the inner more oxidised zone while iron precipitates in a more reduced zone (Rott, 1985).

Manganese can form mixed valency oxides (Fig. A.6) however the conditions required for precipitation are not solely due to redox but are influenced by the presence of other manganese oxide surfaces for catalysing the reaction (Hem, 1989; Knocke *et al.*, 1990). Bacteria would also be a significant catalyst under the conditions found in many groundwaters. Iron may often be found coprecipitated with the manganese oxides as well as other metals scavenged from the water.

APPENDIX B: WELL FOULING DEPOSIT COMPOSITION

B1 Organic

These deposits are likely to be found where there is very high nutrient loading and organic supply for bacterial growth. The main environment where this may be a problem is near injection points where the goal is to stimulate biological activity through the addition of growth factors (e.g. nutrients, oxygen) that may be limiting. Little data is available in the literature on this type of deposit.

B2 Organo-mineral

The main type of deposit occurring in water supply wells is organo-mineral type deposits.

These contain a complex assemblage of materials that can be grouped into:
- Mineral matter
- Organic matter
- Dissolved salts trapped within the deposit

Mineral matter may consist of chemical and microbially precipitated elements (e.g. Fe, Al, S), adsorbed compounds and particles (e.g. Si, clay). Organic matter may comprise a significant component of the dried biofouling deposit ($<70\%$) by volume however when analysed by weight the mineral matter will be the dominant component. The organic matter consists of bacterial cells and bacterially secreted ECP. Dissolved salts trapped within the deposit voids can contribute a significant component of certain elements (e.g. Na). This is significant only in saline groundwater and occurs because 85–97% of the wet weight of the biofouling deposit is residual groundwater. These deposits consist of a skeleton of mineral and organic matter with groundwater filling, may be of the voids within the structure. When a biofouling deposit is scrapped from the surface of the casing between 85% and 97% of its weight is due to water.

These deposits can be classified based on the most dominant mineral or element occurring when the deposits are chemically analysed.

B2.1 Iron Biofouling

The most commonly reported water well fouling deposits are those containing iron, where the biofilms are a mixture of bacteria, iron hydroxide and other associated inorganic precipitates, trapped in a bacterially produced ECP matrix (Plate 14). In an extensive investigation of Australian biofouling deposits (McLaughlan *et al.*, 1993) it was found that under a light microscope the deposits could be classified according to dominant bacteria present. Some biofilms were dominated by the stalked bacteria such as *Gallionella* sp. (Plate 15) while other biofilms mainly comprised heterotrophic bacteria. It was noted that the stalked bacteria dominant biofilms were associated with a lower amount of organic material the deposit.

The composition by weight of typical iron biofouling deposits from a pump in shallow saline aquifers are:
- Wet sample (as scraped from casing)
    - Water, 85–97%
    - Organic matter/carbonates, 1–5%
    - Other elements (e.g. Fe, Na, Ca), 2–10%
- Dried sample (at 105°C for 1 h)
    - Bacterial organic matter ('slime'), 10–18%
    - Carbonates, 5–15%
    - Iron hydroxide, 40–70%
    - Salt (NaCl), 3–13%
    - Clay (SiO_2), 3–10%

The wet sample consists mainly of water while iron is a major component of the dried deposit by weight. SiO_2 is from clay/silt that gets trapped as it passes across the deposit surface although some silica may complex with iron. The salt component is due to the saline water trapped within the deposit upon drying. In less saline water the amount of salt and carbonates would be lower.

B2.2 Aluminium Biofouling

Deposits containing aluminium have been reported from aquatic environments containing acid mine waters, acid sulphate soils and acid geothermal waters (Nordstrom, 1982). In groundwater wells there have been few reports of deposits containing significant amounts of aluminium.

Three types of well fouling deposits containing aluminium have been identified (McLaughlan *et al.*, 1993):
1. Aluminium hydroxide deposits with a sulphate component. Section 6.9 describes a site where this occurred. The deposits were loosely adhered and formed in the well, piping (Plate 13) and the discharge area.
2. Aluminium dominant heterotrophic biofilm. These deposits occurred in a wellfield (Section 6.7) and were generally fibrous and brittle. The deposits that grew on the vertical bars of the well screen had been sculptured by flow through the screen (Plate 12). The deposits had a very high amount of organic matter (43–61% at LOI_{400}) present compared with iron biofouling deposits (6–28% at LOI_{400}).
3. Aluminium bound predominantly in an iron biofouling deposit.

B3 Mineral Scale

The mineral scales found in water wells are most commonly a carbonate scale and occasionally a sulphate bearing scale, however mineral scale occurrence is much less widespread than iron biofouling deposits.

B3.1 Carbonate

Calcium carbonate is often reported while the magnesium bearing carbonates (e.g. dolomite, magnesite) are not often found. Although the chemical conditions may be suitable (thermodynamically) for precipitation of the magnesium bearing carbonates other factors (e.g. kinetics) can inhibit their crystallisation at ordinary temperatures (Deer *et al.*, 1966). These deposits are hard and range in colour from white (Plate 11) to red-black (Plate 16) depending on the amount of iron present as an impurity. The hydrogeological environments in which they occur is varied. At one site an upper saline aquifer corroded the upper casing leading to massive deposition. Around the Toowoomba region in Australia where carbonate fouling occurs the aquifers include basalt, sandstone, shale and alluvial strata.

B3.2 Sulphate

Scales comprising calcium, barium or strontium with sulphate are more often found in wells used for oil and gas production. During operations to increase well production water injection/flooding may be used. The use of a high sulphate water (e.g. seawater) for injection can react with high salinity groundwater to form these scales. Extensive scaling of hydrocarbon recovery pumps by a calcium sulphate scale containing NaCl with some iron hydroxides and organic matter has been reported (Hodder & Peck, 1992).

APPENDIX C: CORROSION BACKGROUND

C1 Electrode Reactions

The reactions involved in electron transfer are oxidation–reduction reactions. An oxidation reaction involves an increase in valence or the extraction of an electron from an ion or atom. The location where this occurs is called the anodic area. Reduction reactions involve a decrease in valence or the consumption of electrons and occur in a cathodic region.

For corrosion to occur there must be the formation of ions and release of electrons simultaneously and at an equivalent rate to the acceptance of electrons at the cathode. The corrosion will then occur at the anodic area. For a corrosion cell to exist the anodic and cathodic areas must be connected so that an electrical circuit is formed. In groundwater wells, water with dissolved ions will allow the transfer of ions to the anodic and cathodic sites while metal surfaces will facilitate electron transfer.

C1.1 Anode

At the anode positively charged metal atoms leave the solid surface and go into solution. The oxidation reaction occurring may be generalised as:

$$M \leftrightarrow M^{n+} + ne$$

where
M = metal
e = electron
n = number

The value of n depends primarily on the nature of the metal.

C1.2 Cathode

At the cathode, electrons generated from the anode react with positive ions in the water to preserve solution neutrality.

Cathode reactions that can occur during the corrosion of metals are (NACE, 1984):
- Oxygen reduction (acid solution)
  $$O_2 + 4H^+ + 4e \leftrightarrow 2H_2O$$

- Oxygen reduction (neutral and alkaline solution)
  $$O_2 + 2H_2O + 4e \leftrightarrow 4OH^-$$
- Hydrogen evolution
  $$2H^+ + 2e \leftrightarrow H_2$$
- Metal ion reduction
  $$M^{n+} + e \leftrightarrow M^{(n-1)+}$$
- Metal deposition
  $$M^{n+} + ne \leftrightarrow M$$

At the cathode more than one of these reactions may occur simultaneously.

C2 Passivity

Metals may become passive and lose their chemical reactivity. This passivity may be due to a thin oxide or adsorbed layer protecting the metal surface. If the passive film is disrupted strong corrosion cells may operate between the active and passive surfaces of the metal. The concentration of dissolved oxygen or other strong oxidiser can significantly affect the corrosion potential of a passivating metal such as Cr–Ni stainless steel.

C3 Polarisation

The retardation of electrochemical reactions (polarisation) due to protective films may result from a buildup of corrosion products (e.g. $Fe(OH)_3$). Other ions (Ca^{2+}, Mg^{2+}) may also precipitate and polarise the surface. These precipitates will decrease the diffusion of O_2 and H_2 at the metal surface and decrease the corrosion rate. Depolarisation that will increase the corrosion rate can also occur from sulphate reducing bacteria utilising the hydrogen at the cathode.

C4 Concentration Cell Formation

A concentration cell can form due to the presence of a deposit causing different chemical conditions (particularly from oxygen) at two sites on the surface of a metal (Fig. C.1). This will lead to differences in the potential between the sites and corrosion.

C5 Galvanic Series

Galvanic activity occurs when a metal is electronically connected to a dissimilar metal in the presence of an electrolyte. This is a type of electrolytic corrosion.

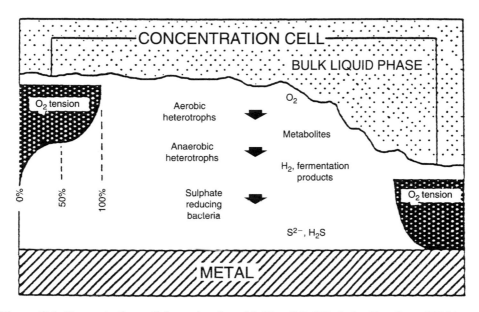

Figure C.1 Concentration cell formed under a biofilm. (Modified after Hamilton, 1985.)

Table C.1 Simplified galvanic series.

Noble (cathodic)	Stainless steel (Type 316, passive)
	Stainless steel (Type 304, passive)
	Copper–nickel alloys
	Bronzes
	Brasses
	Cast iron
	Mild steel
	Zinc
Active (anodic)	Magnesium

The potential of a metal in solution is related to the energy that is released when the metal corrodes. The corrosion potential of metals can be used to rank them relative to each other in a galvanic series (Table C.1). The groups refer to materials that have similar properties and galvanic corrosion is unlikely if materials within the same group are electrically connected. Even though the potential may change with ionic concentration, degree of aeration and temperature of the solution, in general the relative position of the metals will be similar in most environments (NACE, 1984).

The degree of galvanic corrosion will be determined in part by the potential difference between the metals and the ratio of cathodic to anodic area. As the ratio of cathodic to anodic area increases the corrosion rate of the anode increases.

C6 Environmental Factors

The environment around a metal surface will control the corrosion rate. This includes both the chemical and physical properties of the water:

- Salinity

The corrosion rate generally increases as the salinity of the water increases. The high conductivity water will increase the current transfer between the different sections of the casing or between the casing and the earth. Some metal salts (e.g. Cl^-, SO_4^{2-}) hydrolyse in water forming acids (e.g. HCl, H_2SO_4) and are involved in pitting corrosion, while other ions such as bicarbonate, carbonate and hydroxide may decrease the corrosion rate by forming protective scales.

- Oxygen

The dissolved oxygen content is recognised as the major factor influencing corrosion in most wells. The oxygen depolarises the cathode, which increases corrosion. The amount of oxygen supplied to the metal surface is a function of the water temperature, flow rate and the presence of a scale.

- Carbon dioxide

Carbon dioxide is soluble in water and reacts to form a weak acid (H_2CO_3). Corrosion by carbon dioxide occurs in groundwater wells particularly in deep wells where its effect is exacerbated by elevated water temperatures. The corrosion is often very localised in the form of pits, gutters or attached areas with abrupt changes from corroded to non-corroded areas.

- Hydrogen sulphide

Hydrogen sulphide is recognised as a corrosive agent in the natural gas industry where the water has high temperatures, pressures and chloride content. These conditions are unlikely to occur in many water supply wells. However the sulphide inclusions particularly in older casing when quality control procedures were not rigorous is possible. These can act as sources of pit nucleation.

- Flow rate

Flow dependent corrosion may be classified as chemical or mechanical (erosion). In general, corrosion rates increase with increasing velocities up to a point. This reflects the increased reductant (e.g. oxygen, hydrogen) supply to the reaction surface (cathode). At high velocities any scale present can become protective of the surface.

- pH

The relationship between pH and corrosion rate reflects a mixture of hydrogen ion (H^+) effects and carbonate equilibria processes. The increased H^+ concentration that occurs at a low pH accelerates the corrosion of most metals. At a high pH there are more carbonate and hydroxide ions which can increase the tendency of the water to form scales which protect the metal surfaces.

- Temperature

Elevated water temperatures in conjunction with other environmental factors to increase corrosion rates.

APPENDIX D: FIELD CORROSION STUDY

Since the corrosion rate of metals depends on numerous factors (e.g. water chemistry, metallurgy and surface condition of alloys, flow rate, temperature), a rigorous experimental programme proves extremely complex. The corrosion test programme outlined was designed to evaluate a number of these factors in the field.

The sites were carefully selected to sample a wide range of water quality types and well operating schedules under conditions representative of submerged well screen and well casing environments. The experiment was based on determining the weight loss of a range of ferrous-base metals typically used in well construction over a 12-month period. From an evaluation of weight-loss change with time, corrosion rates (the slope of the weight loss versus time curve) for the different metals were related to water quality and other variables. The full reporting of the methodology and results are contained in McLaughlan *et al.* (1993) and summarised in McLaughlan & Kelly (1994).

D1 Methodology

D1.1 Corrosion Device Details

The corrosion device consisted of 80 mm Class 12 uPVC tube with four separate bolts as holders of seven metal coupons mounted within it. To minimise turbulence, alternate bolts were located at right angles, and the length of the inlet side of the device was greater than on the discharge side. In the design of the test devices, particular care was taken to ensure that all coupons were electrically insulated from the bolt using a sleeve of plastic tubing and from each other by small PVC spacers which also uniformly positioned the coupons across the flow path.

At selected time intervals one bolt and coupon assembly was removed from the device for determination of coupon weight loss and a replacement bolt (sleeved but with no coupons) installed to seal the device for the remainder of the test.

D1.2 Corrosion Coupons

The metals selected (Table D.1) are commonly used in groundwater well casing, screens and pump column. They were set up to be evaluated both as an individual metal and as a specific galvanic or dissimilar metal couple. In metallurgical terms,

Table D.1 Coupon details.

Item no.	Material	Thickness (mm)	Finish faces	Edges
1	Galvanised steel	2	Degrease only, mill scale to 220 grit	As cut
2	304SS/MS/304SS	2 + 3 + 2	SS – Std. 2B MS – Polished to 220 grit	As cut Polished to 220 grit
3	3Cr12/MS/3Cr12	2 + 3 + 2	SS – Std. 2B MS – Polished to 220 grit	As cut Polished to 220 grit
4	MS	3	Uniform mill scale	Polished to 220 grit
5	316SS	2	Std. 2B	As cut
6	3Cr12	2	Std. 2B	As cut
7	304SS	2	Std. 2B	As cut

Galvanised steel – Lysaught Zincform sheet G300 and Z600 coating class.
MS – mild steel; SS – stainless steel.
MS – K1022 or equivalent merchant bar hot rolled square edge flat.
304SS – cold rolled strip, 2B finish; 316SS – cold rolled strip, 2B finish; 3Cr12 – cold rolled strip, 2B finish.

stainless alloy 3Cr12 is a ferritic steel similar to Type 409, whilst Types 304 and 316 are of austenitic structure.

The single metal type coupons (Table D.1, items 1, 4, 5, 6 and 7) were 32 mm square section, and after assembly, an exposed area (two faces and edges) of 21.6 cm^2 (item 4) and 20.7 cm^2 (items 1, 5, 6 and 7) remained. For the dissimilar metal coupled coupons (items 2 and 3) the alloy steel section was only 24 mm square; this was spot welded to the mild steel coupon at four points to ensure galvanic connection for the duration of the test. This gave an exposed area of 12.8 cm^2 for the steel and 11.2 cm^2 for the alloy steel on the dissimilar metal coupled coupons.

After machining and numbering the coupons (comprising 672 items), they were degreased with redistilled acetone, weighed and then packed in sealed bags with dried silica gel. The corrosion devices were assembled using gloves to avoid contamination of the metal surfaces and then shipped to the sites, where site personnel installed them.

D1.3 Device Location Details

The experiment comprised setting up 24 devices at 21 sites, with the typical water chemistry data, operating conditions and corrosion indices are given in Table D.2. There was a selection of sites with known corrosivity problems such as high salinity (sites E and F), high CO_2 (sites H, J, K, Y and Z) and low pH (sites P, Q and R) waters. The other sites were selected as typical of town water supply quality. This data set covers a range of water qualities of most concern to the managers of water supply or salinity monitoring schemes.

Table D.2 Physical and chemical data at corrosion sites.

Device	pH	Temp °C	EC mS/m	TDS mg/L	HCO$_3$ mg/L	Cl mg/L	CO$_2$ mg/L	O$_2$ mg/L	Hardness mg/L CaCO$_3$	Alkalinity mg/L CaCO$_3$	SO$_4$ mg/L	Langelier Index	HCO$_3$/Cl	Velocity m/s	Flow L/s	Pump cycle hours on (off)
A	5.3	20	23	146	14	57	—	0.2	18	11	5	−5	0.246	0.03	6	10(14)
B	6.9	25	41	—	155	43	—	0.7	140	—	5	−1	3.605	0.03	6	24(0)
C	5.9	19	37	—	48	71	—	0.5	66	—	25	−3	0.676	0.03	6	24(0)
D	6.4	27	154	819	57	400	—	0.5	114	—	50	−2.5	0.143	0.03	6	—
E	6.9	18	2630	—	165	8530	11	4.3	—	—	1250	0.25	0.019	0.015	3	12(12)
F	6.8	18	1800	—	221	5750	30	0.3	—	—	880	−0.26	0.038	0.03	6	24(0)
H	6.5	35	85	500	105	200	100	—	—	—	—	−1.3	0.525	0.015	3	24(0)
J	6.7	48	62	370	155	125	55	—	—	—	—	−0.8	1.240	0.025	5	24(0)
K	5.7	23	57	275	8	130	19	0.7	28	—	15	−4.4	0.062	0.03	6	24(0)
L	7.3	15	135	740	378	226	—	—	372	310	24	0.25	1.673	0.03	6	8(16)
M	6.6	41	37	270	96	65	—	—	79	65	22	—	1.477	0.03?	6?	24(0)
N	6.8	41	31	270	145	32	—	—	120	70	12	—	4.531	0.03?	6?	24(0)
P	4.4	25	57	300	3	175	290	4	38	2	5	−6.5	0.017	0.044	8.7	24(0)
Q	4.4	25	57	300	3	175	290	4	38	2	5	−6.5	0.017	0.048	9.6	24(0)
R	4.4	25	57	300	3	175	290	4	38	2	5	−6.5	0.017	0.018	3.5	24(0)
S	5.3	25?	114	520	—	320	180	—	270	2	28	—	—	0.05	10.8	On demand
T	7	22	134	—	389	244	—	0.6	—	320	11	−0.3	1.594	0.006	1.2	8(16)
U	6.2	25?	73	—	88	113	90	4.2	—	72	26	−2	0.779	0.009	1.7	6(18)
V	7	23	205	—	693	365	80	0.6	—	525	25	0	1.899	—	—	20(4)
W	6.9	28	76	310	199	47	—	4.69	139	163	33	−0.6	4.234	0.03	6	24(0)
X	7.3	30	72	395	218	44	—	4.43	109	179	43	−0.3	4.955	0.01	2	24(0)
Y	5.2	30	20	—	6	8	50	—	5	—	—	−5.9	0.750	0.03	6	On demand
Z	5.2	30	20	—	6	8	50	—	5	—	—	−5.9	0.750	0	0	On demand
AA	7.3	30	29	160	178	4	—	3.26	149	146	—	—	—	0	0	—

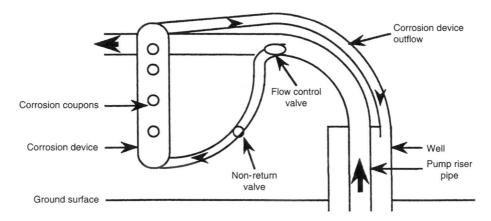

Figure D.1 Typical installation of wellhead corrosion device.

D1.4 Device Installation

At sites Y and AA which were under non-flowing conditions they were suspended by wire to the wellhead and located near the screens. At all other sites there was flow through the device at the wellhead in a similar configuration to Figure D.1. A control valve was supplied to regulate the flow through the corrosion device. A non-return valve was supplied for the inlet line so that backflushing and aeration of the coupons would not occur during periods when the pump was not operating.

The devices placed under flowing conditions were generally scheduled to have a flow of 6 L/s that corresponds to a velocity past the coupons of 0.03 m/s. This velocity is recommended in the well design literature for screen entrance velocities (Driscoll, 1986). Several devices were scheduled to have a higher and lower rate to evaluate the effect of flow rate on corrosion rate.

Unfortunately it was not always possible to achieve the design flow rate. Some wells were also operated intermittently and the coupons were subjected to fluctuating flow conditions. Although this is more realistic of well operating environments it makes the comparison of the data more difficult.

At specified time intervals the coupon bolt assemblies were removed and analysed. Each coupon was prepared by brushing the surface clean of rust, drying at 105°C and then weighing.

D2 Conclusions

This field based corrosion experiment has highlighted the complexity of interplaying factors in determining long-term corrosion rates. The most useful way to utilise this data for planning purpose would be to select a site of similar water quality and pump operating schedules and then calculate the corrosion rate (Figs D.2–D.4). The rate of corrosion loss with time often decreased over time due to the buildup of a protective corrosion product. But this was not always the case and the onset of surface polarisation varied from 70 days to at least 400 days. This is clearly in line with the well-known problem of using short-term corrosion rate data for well design since the corrosion rate will be markedly greater than the actual rate. When using this data for design purposes the wells that have an exponentially decreasing corrosion rate will yield a more realistic long-term corrosion rate.

The corrosion rate of galvanised and mild steel coupons were generally similar except in the wells that had lower corrosion rates where the galvanised steel corrosion rate was lower. This shows that in corrosive waters where the galvanised coating is consumed that galvanising offers no added protection over mild steel.

There generally was no corrosion of the Type 304 or 316 stainless steels during the test. The 3Cr12 alloy steel suffered significant localised corrosion that may make it unsuitable for many applications in groundwater wells.

The galvanic coupling of mild steel with alloy steels significantly increased the corrosion rate compared with a single metal. The corrosion rate for mild steel was around 60% that for a galvanic couple of mild and stainless steels.

The corrosion rates of coupons in wells which were not pumped were 40–70% that of the wells which are intermittently in use. Increasing the flow rate by a factor of 3 (0.018–0.048 m/s) was found to increase the corrosion rate by 20–40%. A comparison of the data from this study with a previous study (Kelly & Kemp, 1974) showed a corrosion rate in this study of between 3 and 19 times greater over a 6-month period on the same well. The current study used a 24-h pump schedule with a flow rate of 0.035 m/s while Kelly and Kemp had an 8-h operating schedule with a flow rate of less than 0.007 m/s. This indicates the significant effect flow rate and pump operating schedules can have on the corrosion rate for a particular water chemistry.

Sites with high corrosion rates typically had the following water chemistries; water rich in CO_2 with elevated temperature (sites H, J and K) and low pH (4.4) water (sites P, Q and R). All of the high corrosivity sites had a 24-h pump cycle that would have increased the corrosion rate.

Sites with a low corrosion rate (sites U, T, V and AA) had a fairly neutral pH water (6.3–7.3) but they also had a low flow rate and daily pumping duration. At site AA the coupons were under stagnant conditions and a lower corrosion rate would be expected.

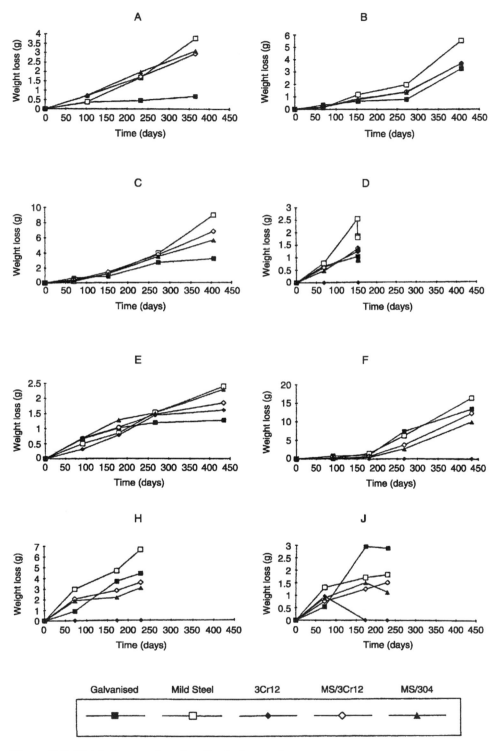

Figure D.2 Weight loss against time for devices A–J.

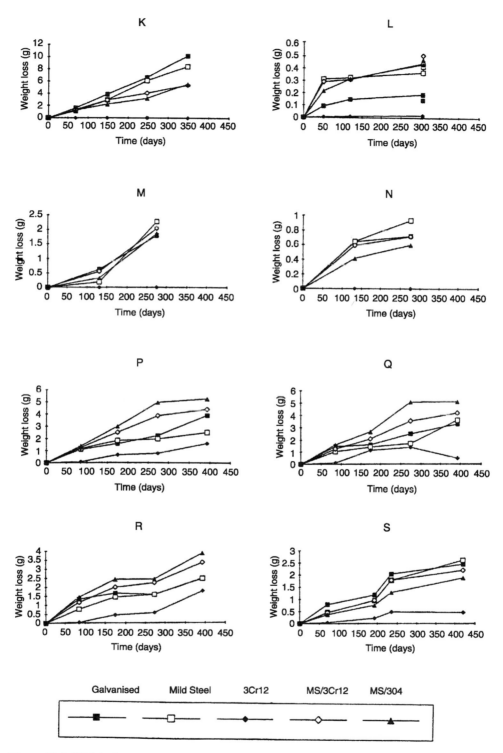

Figure D.3 Weight loss against time for devices K–S.

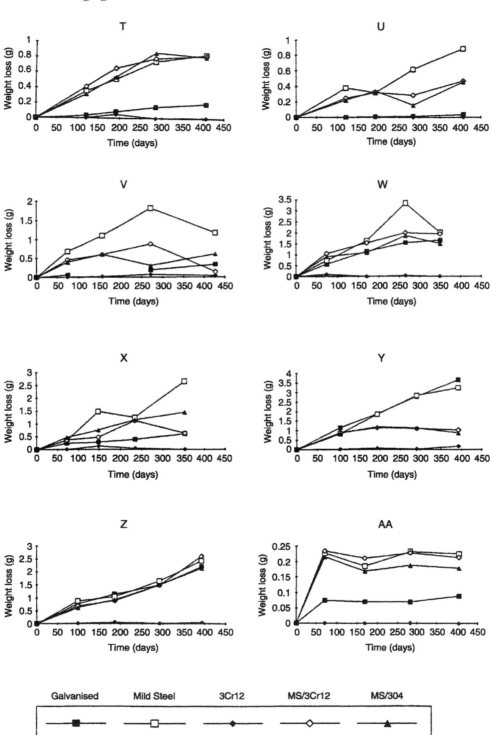

Figure D.4 Weight loss against time for devices T–AA.

APPENDIX E: SAMPLE COLLECTION

E1 Water Samples

Water samples need to be collected, preserved and analysed differently depending upon the parameter to be measured. General water quality parameters such as pH, Eh, EC and dissolved oxygen can be measured in the field using portable instruments. Field chemical sampling kits (e.g. Merck, Hach) are available for iron and manganese concentrations. Details on the collection and preservation techniques for all water quality parameters can be obtained from Standard Methods (APHA, AWWA & WCPF, 1992) or the chemical laboratory where the samples will be processed. The main consideration during water sampling is to ensure that the sample is representative of the water quality occurring within the well. The most common problems are sample aeration and sample contamination during collection.

Some water quality parameters such as EC are relatively insensitive to aeration and may be collected and measured in an open container. Other parameters such as pH are moderately insensitive unless the geology includes carbonate and there is a lot of carbon dioxide present in the water. In this case as the carbon dioxide escapes from the water the pH will rise. This degassing is more of a problem when interpreting laboratory pH measurements where 1–2 weeks may have elapsed between sampling and measurement. It is good practice to collect a field pH measurement for comparison with laboratory measurements. Parameters that are very sensitive to aeration include dissolved oxygen, redox and iron (Fe^{2+}). These samples need to be measured or collected so that air contact is minimised. A simple flow cell to achieve this may involve placing the sample hose into the bottom of an open container with the electrodes suspended so that there is a continual slow flow of water past the electrodes and then over the side of the vessel. This will ensure the electrodes equilibrate with freshly pumped water rather than the oxygenated water at the top of the container, which has been in contact with air. Due to the difficulty in accurately measuring Fe^{2+} it is usually valid to assume that the laboratory measurement of total iron in a water sample is the correct value for Fe^{2+} if there are no colloidal iron hydroxide particles present in the water. If the sample has been collected without turbidity the amount of dissolved Fe^{3+} in normal groundwater is minimal. The oxidation rate of manganese is much slower than iron and hence less susceptible to change during sampling.

Sample contamination is particularly important consideration if sampling from a discharge point on a pumping well. The most significant source of contamination is the flushing of the well that occurs during pump startup. When the well is turned off the water chemistry and microbiology adjust to the tranquil conditions. During pump startup there is a sudden increase in water flow across the well casing especially if the pressure head in the riser pipe needs to be established. This can cause significant changes in the water chemistry during this period. The water chemistry and micro-biology was found to change rapidly during the first 10 min of pumping and then reach a quasi-equilibrium. This indicates that during pump startup the conditions are not typical of those occurring during the majority of the time the pump is spent operating. A sample taken at the wellhead is not necessarily typical of the different types of waters, which occur within the aquifer. It represents a flow-weighted average of all the different water types that enter the well screen. The bulk of the aquifer can be signifi-cantly different from a small iron rich layer that can cause biofouling. The long-term sand production rate should be measured after the pump has been operating for several hours. A surge of sand also may occur immediately after pump startup with the concentrations often decreasing over 15 min of pumping.

E2 Solid Sample Collection

Fouling deposits are not uniformly distributed throughout a groundwater extraction system and will tend to collect in specific places within the system depending upon the type of deposit. Biofouling deposits are concentrated in various parts of the system (Fig. 2.4) while particulate deposits will collect at the bottom of the well and other low velocity zones. An attempt should be made to sample the deposit as close to the source as possible to determine the true nature of the problem. This is particularly important with particle deposits. The particle size of the deposit reflects the minimum energy available to transport the sediment between that point and the source. So a sample of fine-grained material from a setting tank may suggest that only fine material is enter-ing the well possibly because of incorrect gravel pack selection. However if coarse-grained material was found accumulated at the bottom of the well there may be a corrosion hole in the casing. This would allow both fine- and coarse-grained material to enter the well with the finer material in the settling tank transported further through the groundwater extraction system due to its smaller size.

Where direct access to the fouled sections is possible the soft deposits can be easily removed using a knife or spoon, however hard deposit may need chiseling, mechanical shock or heating. Contact with foreign material (i.e. grease, oil) should be avoided during sample collection. Samples for microbiological analyses should be collected using sterile instruments and containers. Water samples for microbial culturing require analysis within 24 h and should be kept refrigerated during transport and stor-age. Refrigeration at 4°C will minimise bacterial activity before analysis. Those water and solid samples for microbial visual inspection may be held for a much longer time frame before analysis since the bacterial structures used for identification will still be apparent. Some liquid from near the deposit should be included in the sample con-tainer if the sample came from a submerged environment. This will preserve as much as possible the environmental conditions under which the deposit was formed. If there is variability in the samples from different parts of the groundwater extraction system then individual rather than mixed samples should be collected and analysed.

Samples from the well casing can be collected using a modified bailer. A simple hand bailer can consist of a piece of tube flanged out at the top with mesh on the opposite end. When lowered on a cable and scraped along the well screen, the biofouling deposit gets trapped in the mesh while the water can pass through. To determine the location and grain size of sand/gravel particles entering the well a bailer with a closed bottom end can be used. The bailer is lowered down the well to the desired location and the well is agitated with a surge block to encourage particle entry into the well. By repeating this procedure from the bottom of the well upwards the section of damaged casing can be located.

Where there is no direct access to the fouled surfaces then solid samples may be filtered/cultured on prepared surfaces placed in contact with the groundwater. Sampling devices can either be lowered down the well under the water level or connected up to the wellhead. The time required to collect an adequate quantity of sample by these devices depends on the degree of filtering involved in the method chosen. The filter sampling method may require 1 h while the coupon devices that rely solely on bacterial growth rather than filtering could be in excess of one week.

APPENDIX F: GLOSSARY

Aerobic   Oxygen is present

Alkalinity   Capacity of water to neutralise acids

Alluvial   Deposited from water

Anaerobic   Oxygen is not present

Anion   An ion which has a negative charge

Anode   Location where oxidation and corrosion occur

Aquifer   Strata which yield sufficient water

Autotrophic   Obtaining energy from inorganic material

Biocide   Chemical which kills biological matter

Biofilm   Biological layer attached to a surface

Biofouling   Growth of biofilm

Cathode   Location where reduction and no corrosion occurs

Cation   Positively charged ion

Cavitation   Formation and collapse of gas bubble

CCTV   Closed circuit television camera

Chelate   Organic compound which bonds another to a metal in solution

Complex   Compound formed by chelation

Cone of influence   Extent of drawdown around a pumping well

Connate   Naturally occurring

Drawdown   Distance between the static water level and the groundwater table or potentiometric surface caused by pumping

ECP   Extracellular polymers produced by bacteria

Eh   Measure of the electrochemical potential

Electrochemical reaction   Chemical reaction involving loss or gain of electrons

Electrolyte   An ionic conductor

Fines   Clay to silt sized particles

Heterotrophic   Unable to obtain energy from inorganic material

Homogeneous   Uniform

Igneous   Rocks formed from molten material

Interstitial   Occurring within the pore space

Ion   An element or compound that has lost or gained an electron

Kinetic   Rate at which a reaction occurs

LSI   Langelier Saturation Index

MIC   Microbially influenced corrosion

Microaerobic   Low concentration of oxygen present

Oxidation   Loss of electrons

Packer   An inflatable device

Passivation   Metal becomes more noble and resistive to corrosion

Permeability   Capacity of aquifer to transmit fluid

pH   Measure of the acidity or alkalinity of a solution

Polarisation   Reduction in the potential of a surface due to a protective film

Redox   See Eh

Reduction   Loss of electrons

RI   Ryznars Index

Riser pipe   Piping used to connect pump/pump inlet to the surface

Sedimentary   See Alluvial

Solubility   Capacity of a substance to dissolve in another substance

Specific capacity   The rate of discharge per unit of drawdown

Strata   Rock layer

Static water level   Water level measured when there is no withdrawal of groundwater

Surfactant   Substance capable of reducing the surface tension of a liquid it is dissolved in

TEAP   Terminal electron acceptor process

Thermodynamically   Occurs based on energy considerations

Unconsolidated   Strata which are not cemented

Valence   Charge associated with an ion

Yield   Discharge rate from well

BIBLIOGRAPHY

For further reading a number of publications are recommended which contain detailed information on the topics listed.

WELL DETERIORATION: GENERAL

ASCE 1996. Operation and maintenance of ground water facilities. In Lloyd C. Fowler (ed.), *Committee on ground water of the irrigation and drainage division*. New York: American Society of Civil Engineers.

Gass, T.E., Bennett, T.W., Miller, J. & Miller, R. 1980. *Manual of water well maintenance and rehabilitation technology*. National Water Well Association.

Howsam, P. (ed.) 1990. *Water wells: Monitoring, maintenance and rehabilitation*. London: E & F Spon.

Howsam, P., Misstear, B. & Jones, C. 1995. *Monitoring, maintenance and rehabilitation of water supply boreholes*. London: Construction Industry Research and Information Association.

Mansuy, N. 1999. *Water well rehabilitation: A practical guide to understanding well problems and solutions*. Boca Raton, FL: CRC Press.

McLaughlan, R.G., Knight, M.J. & Stuetz, R.M. 1993. *Fouling and corrosion of groundwater wells: A research study*. Research Publication 1/93, National Centre for Groundwater Management, University of Technology, Sydney, 213pp.

Smith, S.A. 1995. *Monitoring and remediation wells: Problem prevention, maintenance and rehabilitation*. New York: CRC Press.

WATER WELL REHABILITATION TECHNIQUES/DRILLING

ADITC 1992. *Australian drilling manual*. Australian Drilling Industry Training Committee Ltd, MacQuarie Fields, NSW.

ARMCANZ 1996. *Minimum construction requirements for water bores in Australia*. Agriculture and Resource Management Council of Australia and New Zealand.

Borch, M.A., Smith, S.A. & Noble, L.N. 1993. *Evaluation and restoration of water supply wells*. AWWA Research Foundation, AWWA, Denver, CO.

Driscoll, F. 1986. *Groundwater and wells.* St Paul, MN: Johnson Division.

Roscoe-Moss 1990. *Handbook of ground water development.* New York: John Wiley & Sons.

CORROSION

AWWA 1986. *Corrosion control for operators.* AWWA, Denver, CO.

AWWA-DVGW 1986. *Internal corrosion of water distribution pipelines.* AWWA Research Foundation, Denver, CO.

Ireland, J.B. 1978. *Failure and aging of water well casing and screen by compressive rupture, encrustation and corrosion.* Roscoe-Moss Company, LA.

Jones, L.W. 1988. *Corrosion and water technology for petroleum producers.* Tulsa: Oil and Gas Consultants Inc. (OGCI) Publications.

Roscoe-Moss 1990. *Handbook of ground water development.* New York: John Wiley & Sons.

WELL COMPONENT PERFORMANCE EVALUATION

Helweg, O.J., Scott, V.H. & Scalmanini, J.C. 1983. *Improving well and pump efficiency.* American Water Works Association, Denver, CO.

Kruseman, G.P. & de Ridder, N.A. 1990. *Analysis and evaluation of pumping test data.* Publication 47, International Institute for Land Reclamation and Improvement, Netherlands, applicable for aquifer analysis only.

Roscoe-Moss 1990. *Handbook of ground water development.* New York: John Wiley & Sons.

CHEMISTRY/MICROBIOLOGY

Cullimore, D.R. 1992. *Practical manual of groundwater microbiology.* London: Lewis Publishers.

Hem, J.D. 1989. *Study and interpretation of the chemical characteristics of natural water.* US Geological Survey Water-Supply Paper 2254.

REFERENCES

ADITC 1992. *Australian drilling manual*. Australian Drilling Industry Training Committee, NSW.

APHA, AWWA & WPCF 1992. *Standard methods for the examination of water and wastewater*. American Water Works Association, Denver, CO.

ASTM Standards D2688 1983. *Standard test methods for corrosivity of water in the absence of heat transfer (weight loss methods)*. ASTM, Philadelphia, PA.

ASTM 631-72 1972. *Standard recommended practices for laboratory immersion corrosion testing of metals*. ASTM, Philadelphia, PA.

ASTM G1-81 1983. *Recommended practice for preparing, cleaning and evaluating corrosion test specimens*. ASTM, Philadelphia, PA.

AWWA 1986. *Corrosion control for operators*. American Water Works Association, Denver, CO.

AWWA 1990. *AWWA standard for water wells, ANSI/AWWA A100*. American Water Works Association, Denver, CO.

Bich, N.N. & Bauman, J. 1995. Pulsed current cathodic protection of well casings. *Materials Performance* April: 17–21.

Bliss, K. 1990. The use of geophysical well logging as an aid to well construction and rehabilitation. In P. Howsam (ed.), *Water wells: Monitoring, maintenance and rehabilitation*. London: E & F Spon.

Borch, M.A., Smith, S.A. & Noble, L.N. 1993. *Evaluation and restoration of water supply wells*. AWWA Research Foundation, AWWA, Denver, CO.

Braester, C. & Martinell, R. 1988. The vyredox and nitredox methods of in situ treatment of groundwater. *Water Science and Technology* 20(3): 149–163.

Clarke, L. 1977. The analysis and planning of step drawdown tests. *Quarterly Journal of Engineering Geology* 10: 125–143.

Clarke, L., Radini, M. & Bison, P.L. 1988. Borehole restoration methods and their evaluation by step-drawdown tests: The case history of a detailed study in Northern Italy. *Quarterly Journal of Engineering Geology* 21: 315–328.

Cullimore, D.R. 1992. *Practical manual of groundwater microbiology*. London: Lewis Publishers.

Deer, W.A., Howie, R.A. & Zusseman, J. 1966. *Introduction to the rock forming minerals*. London: Longmans, 490pp.

Driscoll, F. 1986. *Groundwater and wells*, 2nd edition. St Paul, MN: Johnson Division.

Dunn, D.S., Bogart, M.B., Brossia, C.S. & Cragnolino, G.A. 2000. Corrosion of iron under alternating wet and dry conditions. *Corrosion* 56(5): 470–481.

EPA 1975. *Manual of water well construction practices*. Environmental Protection Agency, EPA 570/9-75-001.

Forward, P. 1994. Control of iron biofouling in submersible pumps in the Woolpunda salt interception scheme in South Australia. *IAH/IEA Water Down Under '94 Conference Proceedings, Adelaide, November 1994 Preprints,* Vol. 2, Part A, 169–174.

Forward, P. & Ellis, D. 1994. Selection and performance of materials in saline groundwater interception schemes. *Proceedings Corrosion and Prevention '94, Australasian Corrosion Ass'n Inc., South Australia Branch, CAP 94*, Paper 48: 1–9.

Fujita, H., Momose, M. & Pascual, T.V. 1990. Experimental research and development study on water well construction in area where groundwater contains iron and manganese. *Water Supply* 8(3–4, Water Nagoya '89): 402–409.

Gass, T.E., Bennett, T.W., Miller, J. & Miller, R. 1980. *Manual of water well maintenance and rehabilitation technology.* National Water Well Association.

Gurrappa, I. 1996. Electrochemical prevention of corrosion and fouling – A review. *Corrosion Prevention and Control* April: 48–51.

Hamilton, W.A. 1985. Sulphate reducing bacteria and anaerobic corrosion. *Annual Review of Microbiology* 39: 195–217.

Hem, J.D. 1989. *Study and interpretation of the chemical characteristics of natural water*. Washington, D.C.: U.S. Geological Survey Water-Supply Paper 2254.

Hem, J.D. & Robertson, C.E. 1967. *Form and stability of aluminium hydroxide complexes in dilute solutions.* Washington, D.C.: U.S. Geological Survey Water-Supply Paper 1827-A.

Hodder, E.A. & Peck, C.A. 1992. Aquifer restoration system improvement using an acid fluid purge. *Proceedings of the sixth national outdoor action conference, Las Vegas, NV:* 471–482, National Groundwater Association, Dublin, OH.

Howsam, P. & Tyrrell S.F. 1989. Diagnosis and monitoring of biofouling in enclosed flow systems – Experience in groundwater systems. *Biofouling* 1: 343–351.

Howsam, P., Misstear, B. & Jones, C. 1995. *Monitoring, maintenaince and rehabilitation of water supply boreholes.* Construction Industry Research and Information Association, London.

Ireland, J.B. 1978. *Failure and aging of water well casing and screen by compressive rupture, encrustation and corrosion.* Los Angeles, CA: Roscoe-Moss Company.

IWES 1986. *Groundwater occurrence, development and protection.* Water Practice Manual Nr 5, Institution of Water Engineers and Scientists, London.

Janssens, J.G., Pintelon, L., Cotton, A. & Gelders, L. 1996. Development of a framework for the assessment of operation and maintenance (O&M) performance of urban water supply and sanitation. *Water Supply* 14(1): 21–33.

Jones, L.W. 1988. *Corrosion and water technology for petroleum producers.* Tulsa: Oil and Gas Consultants Inc (OGCI) Publications.

Kelly, G.J. & Kemp, R.G. 1974. *The corrosion of groundwater pumping equipment.* Technical paper No. 12, Australian Water Research Advisory Council, Canberra, Australia.

Knocke, W.R., Van Benschoten, J.E., Kearney, M., Soborski, A. & Reckhow, D.A. 1990. *Alternative oxidants for the removal of soluble iron and manganese.* AWWA Research Foundation, Denver, CO.

Kruseman, G.P. & de Ridder, N.A. 1990. *Analysis and evaluation of pumping test data.* Publication 47, International Institute for Land Reclamation and Improvement, Wageningen, The Netherlands.

Little, B.J., Ray, R.I. & Pope, R.K. 2000. Relationship between corrosion and the biological sulfur cycle: A review. *Corrosion* 56(4): 433–443.

Lutters-Czekalla, S. 1990. Lithoautotrophic growth of iron bacterium *Gallionella ferruginea* with thiosulphate or sulphide as energy source. *Archives of Microbiology* 154: 417–421.

McCrae, I.C., Edwards, J.F. & Davis, N. 1973. Utilisation of iron gallate and other organic complexes by bacteria from water supplies. *Applied Microbiology* 25(6): 991–995.

McCaulou, D.R., Jewett, D.G. & Huling, S.G. 1995. *Nonaqueous phase liquids compatibility with materials used in well construction, sampling and remediation.* EPA/540/S-95/503, Robert S. Kerr Environmental Laboratory, Ada, OK.

McDowell-Boyer, L.M., Hunt, J.R. & Sitar, N. 1986. Particle transport through porous media. *Water Resources Research* 22(13): 1901–1921.

McLaughlan, R.G. & Kelly, G.J. 1994. Field corrosion rates of metals in groundwater. *Proceedings Corrosion and Prevention '94, Australasian Corrosion Ass'n Inc., South Australia Branch, CAP 94*, Paper 49, 1–10.

McLaughlan, R.G., Knight, M.J. & Stuetz, R.M. 1993. *Fouling and corrosion of groundwater wells: A research study*. Research Publication 1/93, National Centre for Groundwater Management, University of Technology, Sydney, 213pp.

NACE 1984. *Corrosion basics: An introduction*. Houston, TX: National Association of Corrosion Engineers.

NACE Test Method 1984. *Methods for determining water quality for subsurface injection using membrane filters*. NACE Standard TM-01-73, Item No. 53016. Houston, TX: National Association of Corrosion Engineers.

Nordstrom, D.K. 1982. The effects of sulphate on aluminium concentration in natural waters: Some stability relations in the system Al_2O_3–SO_3–H_2O at 298 K. *Geochemica et Cosmochima Acta* 46: 681–692.

ODA 1993a. *The cost-effectiveness of monitoring and maintenance strategies associated with groundwater abstraction – A methodology for evaluation, Part I – Report*. Overseas Development Administration Project 5478A, Silsoe College, Cranfield University, UK.

ODA 1993b. *The cost-effectiveness of monitoring and maintenance strategies associated with groundwater abstraction – A methodology for evaluation, Part II – Methodology manual*. Overseas Development Administration Project 5478A, Silsoe College, Cranfield University, UK.

ODA 1993c. *The cost-effectiveness of monitoring and maintenance strategies associated with groundwater abstraction – A methodology for evaluation, Part III – Annexes*. Overseas Development Administration Project 5478A, Silsoe College, Cranfield University, UK.

Payne, R.D.G. 1995. *Groundwater recharge and wells: A guide to aquifer storage and recovery*. Boca Raton, FL: CRC Press.

Pelzer, R. & Smith, S.A. 1990. Eucastream suction flow control device: An element for optimisation of flow conditions in wells. In P. Howsam (ed.), *Water wells: Monitoring, maintenance and rehabilitation*: 209–216. London: E & F Spon.

Prevost, R.C. 1987. *Corrosion protection of pipelines conveying water and wastewater: Guidelines*. World Bank Technical Paper No. 69, Water Supply Operations Management Series, Washington, D.C.

Puri, S. & Flores, C.V. 1990. Monitoring and diagnosis for planning borehole rehabilitation—The experience from Lima, Peru. In P. Howsam (ed.) *Water wells: Monitoring, maintenance and rehabilitation*: 377–390. London: E & F Spon.

Puri, S., Petrie, J.L. & Flores, C.V. 1989. The diagnosis of seventy municipal water supply boreholes in Lima, Peru. *Journal of Hydrology* 106: 287–309.

Roscoe-Moss 1990. *Handbook of ground water development*. New York: John Wiley & Sons.

Rott, U. 1985. Physical, chemical and biological aspects of the removal of iron and manganese underground. *Water Supply* 3(2): 143–150.

Schaffner, Jr. R., Lee,Y., Holdaway, B.H., Seeley, R., Mayo, A.L. & Borup, M.B. 1990. Bacterial pore-clogging as a primary factor limiting the enhanced biodegradation of highly contaminated aquifer. *Conference on petroleum hydrocarbons and organic chemicals in groundwater water-prevention, detection and restoration, Houston, Texas: 401–415*. Water Well Journal Publishing Co.

Schippers, J.C., Verdouw, J. & Zweere, G.J. 1995. Predicting the clogging rate of artificial recharge wells. *Journal of Water Supply: Research and Technology—Aqua* 44(1): 18–28.

Semprini, L., Roberts, P., Hopkins, G.D. & McCarty, P.L. 1990. A field evaluation of *in-situ* biodegradation of chlorinated ethenes: Part 2, Results of biostimulation and biotransformation experiments. *Groundwater* 28(5): 715–727.

Smith, S.A. 1992. *Methods for monitoring iron and manganese biofouling in water wells*. AWWA Research Foundation, Denver, CO.

Smith, S.A. 1995. *Monitoring and remediation wells: Problem prevention, maintenance and rehabilitation*. New York: CRC Press.

Tyrrel, S.F. & Howsam, P. 1990. Monitoring and prevention of iron biofouling in groundwater abstraction systems. In P. Howsam (ed.), *Water wells: Monitoring, maintenance and rehabilitation*: 100–106. UK: E & F Spon.

van Beek, C.G.E.M. 1984. Restoring well yield in the Netherlands. *Journal of AWWA* 76(10): 66–72.

van Beek, C.G.E.M. 1989. Rehabilitation of clogged discharge wells in the Netherlands. *Quarterly Journal of Engineering Geology* 22: 75–80.

Water Codex 1982. Washington, D.C.: National Academy Press.

Wilson, B.L. 1990. The effects of abrasives on electrical submersible pumps. *SPE Drilling Engineering* June: 171–175.

Wolf, D.E. & Miller, R.R. 1989. A pilot study for the optimisation of iron control in recovered ground water at a leaking underground storage tank site. *Conference on Petroleum Hydrocarbons and Organic Chemicals in Groundwater Water–Prevention, Detection and Restoration, Houston, Texas: 427–439*. Water Well Journal Publishing Co.

SUBJECT INDEX